BANANAS AND PLANTAINS

NIPA® GENX ELECTRONIC RESOURCES & SOLUTIONS P. LTD.

New Delhi-110 034

BANANAS AND PLANTAINS

Postharvest Management, Storage, Ripening and Processing

C.K. Narayana
Head
Division of Postharvest Technology
Indian Institute of Horticultural Research
Hessarghatta Lake (P.O.)
Bengaluru – 560 089, Karnataka

NIPA® GENX ELECTRONIC RESOURCES & SOLUTIONS P. LTD.
New Delhi-110 034

NIPA® GENX ELECTRONIC RESOURCES & SOLUTIONS P. LTD.

101,103, Vikas Surya Plaza, CU Block
L.S.C. Market, Pitam Pura, New Delhi-110 034
Ph : +91 11 27341616, 27341717, 27341718
E-mail: newindiapublishingagency@gmail.com
www: www.nipabooks.com

For customer assistance, please contact
Phone: + 91-11-27 34 17 17
Fax: + 91-11- 27 34 16 16
E-Mail: feedbacks@nipabooks.com

ISBN: 978-81-19002-84-9

Composed and Designed by NIPA®.

Dedicated to my wife
Lakshmi and son Jayaditya

Indian Institute of Horticultural Research

Hessaraghatta Lake Post, Bangalore – 560 089
Tel. No.91-80-28466420-423, 28446140-143,
Fax: 91-80-28466291

Foreword

Banana is an important crop for India and contributes 30% to total fruit production in the country. It is intricately entwined with the Indian culture and tradition. Cultivation of banana in India is an important activity for livelihood security of small and marginal farmers. It also provides employment to a large number of people involved in its trade and processing. Being a climacteric fruit, it is highly perishable and needs extreme care during its post harvest handling, storage and processing. The post harvest losses in fruits and vegetable is high in India, more so in bananas and plantains. Improper post harvest practices like faulty harvesting method, damage during handling, transportation and retailing results in cuts, wounds, bruises and impact to the fruits which spoils faster due to rapid biochemical changes and microbial attack.

India is the largest producer of banana in the world, but its share in international trade is negligible. However, international trade in banana is one of the largest both in terms of volume and earnings. In international trade, quality of fruits in terms of appearance is very important. The quality is influenced by several abiotic and biotic factors both during as well as after the production. It is very important for the producer and post harvest handlers to understand these in order to produce exportable quality fruits. Banana has the highest processing potential as every part of the plant is suitable for use in one or the other way. Presently banana chips is very popular product marketed all over the world. However, other products like banana fig, candied fruit, flour based products and alcoholic and non-alcoholic beverages from banana also have high potential. It is a crop with highest biomass production and only one-third of it is utilized. Advancement in biomass utilization research has resulted in development of several industrial products from banana waste.

A lot of research has been carried out on various aspects of banana in different parts of the world. This book provides a wealth of updated information on post harvest management and processing of banana and its wastes. The effort of the author to collate and present the information in the form of a book is highly commendable. The book will be a highly useful reference for the researchers, students and common man to understand the developments in science, technology and trade of banana. It provides latest information on post harvest aspects of banana handling, transportation, packaging, storage, processing and waste utilization.

Date : 02.12.2013

Amrik Singh Sidhu
Director

Preface

Fruits and vegetables differ from other food commodities, by its living nature. Though detached from the parent plant, they continue to carry on most of the metabolic processes, notably the catabolic reactions, till it is consumed or processed. The methods of post harvest management determines the quality and shelf life of fresh fruits while the methods of processing employed for conversion determines the retention or enhancement of its nutritive value and the storage life. Banana is the most important fruit for India, due to its year round production and dependence of many farmers and traders on it for their livelihood. India is the largest producer of banana in the world with highest productivity and diversity of clones/varieties/land races. From the point of trade, though India does not have significant place in exports, it has been the largest consumer of bananas consuming almost all that it produced.

After attaining the first place in production, India started valuing the quality (physical and sensory) in all commodities. Though the original quality cannot be improved after the harvest, it can be maintained by adopting suitable handling practices and loss can be reduced through appropriate handling and storage techniques. The authoritative books on bananas mostly covered production and protection aspects while post harvest management and processing was limited to a couple of chapters. Ever since the publication of those, lot of development took place in various areas of post harvest technology of banana including crop residue/waste utilization. The year 2012 witnessed the end of 'Banana Wars', and the international trade in banana found new players and rules. At this juncture when Indian traders are contemplating exports in a big way, the farmers, traders and fruit handlers need to be educated on proper methods of harvest, handling, packaging, transportation, storage and ripening of banana.

This book exclusively deals with every important aspect of post harvest of banana including processing and residue utilization. With heightened health concerns and affordability, health food industry is growing by leaps and bounds. Banana flour is finding a new role in this segment due to its unique qualities and banana leaf in clinical treatment of wounds in humans and animals.

The first few chapters introduces the reader to the crop, its origin and distribution, varieties cultivated through out the country and their characteristics. Later the trade of banana, both international and domestic, is explained along with the ways the fruit is consumed in different parts of the world. A chapter exclusively deals with the nutritive and therapeutic values of banana followed by the post harvest aspects at length in seven chapters with all the latest scientific developments. The last three chapters explains about the processing and value addition including the waste/by-products utilization. The readers will find it comprehensive with all the information relevant to post harvest aspects of bananas and plantains. Constructive criticism and suggestions would be welcomed to improve its content and presentation in future prints.

C.K. Narayana

Contents

Chapter – 1

Introduction

Bananas and plantains constitute the fourth most important global food commodity (after rice, wheat and maize). They are cultivated in over 120 countries in the tropical and subtropical regions of the world in an area of approximately 100 million hectares, where they constitute a major staple food crop for millions of people, as well as providing a valued source of income through local and international trade.

Banana is one of the oldest fruit crops of India, finding its reference even in epics like Ramayana. India is believed to be one of centres of origin, besides southeast Asian countries like Burma, Thailand, Malaysia and Indonesia. It is a crop having multifarious uses as every part of the plant is utilized and hence aptly called as 'Kalpatharu' the plant of virtues. It is important internationally from the point of view of trade as well as livelihood security.

Though the word 'banana' is of Egyptian origin, ('banan' meaning finger) in ancient Indian literature it was referred to as 'Kadali'. In a typical Hindu religious ceremony or ritual, banana occupies a place of prime importance, from birth to death of an individual. Bananas rose from the status of a wild species of the genus *Musa* growing in the jungle to the present, where it is one of the most important internationally traded fruits of the world, which can make or break a nation.

India is the largest producer of banana in the world constituting almost 28% of the total global production. The annual production of bananas and plantains in India is 29.78 million tonnes (NHB, 2012). It is followed by China, the Philippines and Ecuador. Other leading banana producing countries are Brazil, Indonesia, Tanzania, Guatemala, Mexico and Colombia.

Banana belongs to the family Musaceae and genus *Musa*. This genus is further classified into five sections *viz*. Eumusa, Rhodochlamys, Callimusa, Australimusa and Incertae sedis which accommodates all the wild and cultivated species of bananas and plantains. Most of the present day edible bananas are triploids derived from natural crosses of two important species, *Musa acuminata* and *Musa balbisiana*. They are further classified based on their genomic status into AA, AB, AAA, AAB, ABB, AAAA and ABBB. Unlike the corporate banana cultivation in major exporting countries like Ecuador, Costa Rica and Columbia, most of the banana farms in India are small and marginal. Dissimilar to perennial banana plantations of Africa, in India it is grown as an annual crop with a few exceptions of one or two ratoons.

In India, more than 50 cultivars of banana are cultivated, each having its own significance in the area where it is grown, as well as among the consumers and the traders. Bananas are mostly dessert types used for table purpose, while plantains are cooking types or dual purpose varieties. Grand Naine, Robusta, Poovan, Neypoovan, Rasthali, Karpuravalli, Tellachakrakeli, Red Banana, Nendran, Monthan, Batheesa, Hill Banana (Veerupakshi & Sirumalai) are the popular varieties grown all over the country with different local names. Monthan, Nalla Bontha, Karibotha and Batheesa are cooking types and Nendran and its variants are dual purpose varieties.

Banana is one of the cheapest sources of wholesome nutrition to many in the world. It contains nearly all the essential nutrients including minerals and vitamins and has several medicinal properties. Banana is a rich source of energy and is reported that about 24 bananas each weighing around 100gm provides the total energy requirement (2400 κcal/day) of a sedentary man. Besides as a fresh fruit it is also used in beverages, icecreams, confectionary and baby foods. Culinary bananas are also used for making various kinds of traditional foods in various parts of the world. Banana is rich in carbohydrates, minerals like potassium, phosphorus, magnesium and calcium. It has certain phytochemicals like serotonin and dopamine, which are neurotransmitter, responsible for proper functioning of the brain. It is rich in dietary fibres and helpful for easy bowel movement. It has also got several other medicinal and therapeutic values, due to which it extensively used as a food medicine.

Bananas occupy a significant place in international trade of fruits, as it is not grown in temperate regions of the world. Internationally bananas are traded to a tune of 14 million tonnes, valued at more than US$ 5 billion. Though India produces, one third of the total world bananas and plantains, almost all of them are consumed domestically. India's contribution to international trade is only 54,304 tonnes bringing in a foreign exchange of approximately US$ 20 million.

Bananas being climacteric fruit, are harvested when still green, and then ripened using an appropriate ripening aid. Post harvest management of banana is a tricky issue, as its quality and storage life is dependent on several factors. Maturity, harvesting method, packaging, transportation and storage conditions play a vital role in presenting it in an acceptable condition in the market. Besides the variety, manipulation of these factors help in success of banana trade. Many processed products of banana are also popular in different parts of the world. Banana puree, chips, flour, powder and dehydrated banana are some of the important products traded globally. Each country has several other domestically consumed products of banana ethnic to their region.

Two-third of total biomass produced during banana cultivation being non-economical produce, is ploughed back into the field. Among the fruit crops next to pineapple, banana is an important source of fiber. Banana fiber has several applications since time immemorial. Other parts of the plant like leaves, stem and rhizome also have several uses.

Chapter - 2

Origin, Production and Distribution

2.1 Origin

The antiquity of bananas is traced back to 8000 B.C., and its reference in Indian epic, Ramayana (2029 C.C) and pictorial depictions in paintings and sculptures of Ajanta and Ellora testify its existence in India since the early times of Indian civilization. Anthropologists believe that its earliest use was for non-food purpose like fibre, shelter and float to swim across water ways. The earliest bananas were seedy, non-pulpy and non-edible weeds growing along the water ways and jungles. These originally belonged to two diploid species, viz. *Musa acuminata* (AA genome) and *Musa balbisiana* (BB genome). The former was plentifully distributed in South East Asian countries of Thailand, Malaysia, Indonesia and the latter in Indian subcontinent and the Philippines. The diploid AAs gave rise to triploid AAA through a process known as chromosome restitution. Movement of people across these regions brought *Musa acuminata* into Indian subcontinent where it introgressed naturally with the *balbisiana* species, resulting in development of a spectrum of genomic groups like AA, BB, AAA, AAB, ABB, AABB, ABBB, etc. The evolution of triploids is considered to have occurred due to fertilization of a diploid egg with a haploid pollen (Simmonds, 1962). These natural hybrids with various characteristics are likely to have arisen in various ways and locations at various times in the past. A majority of cultivated types are AAAs, triploids which are sweeter dessert types, while AABs and ABBs are starchy cooking types. Triploidy in bananas conferred benefits like vigor, sturdiness and higher productivity, which attracted man to domesticate it. Sterility, due to triploidy, led to vegetative propagation of banana using suckers. Identification of new species like *Musa schizocarpa* and *Musa textiles* and the presence of *S* and *T* genomes in the cultivated and wild types is leading to believe that, besides *acuminata* and *balbisiana* species these two have also been among the progenitors to the present day bananas (Carreel, 1994). In India the banana biodiversity is spread across North Eastern states, Western Ghats, Eastern Ghats and Andaman and Nicobar Islands (Singh and Uma, 2000).

As per the international classification system (Daniells *et al.*, 2001) *Musa* has 5 major sections, viz., *Eumusa, Rhodochlamys, Callimusa, Australimusa* and *Incertae sedis*. There are 11 wild species including *Musa acuminate* and *Musa bulbisiana* in section *Eumusa*, which contribute to 56% of total diversity present in India. The cultivated bananas are grouped into 8 genomic groups viz., AA, AAA, AAAA, AB, AAB, ABB, ABBB, BBB.

Banana is presently cultivated commercially in almost 120 nations across the globe, with a production of close to 100 million tonnes from an area of 4.80 million hectares (NHB, 2011). India is the largest producer of banana with 29.78 million tonnes followed by China, Philippines and Ecuador. The other leading banana producing countries are Brazil, Indonesia, Tanzania, Guatemala, Mexico and Colombia (NHB, 2011).

In India, we find three systems of banana cultivation, viz. irrigated or wetland cultivation, homestead or backyard cultivation and rainfed hill cultivation. There are a large number of cultivars, more than 50, that are commercially cultivated all over India. However, Cavendish bananas called 'Grand Naine' occupy the largest area. Other important cultivars are Robusta, Poovan, Neypoovan or Elakki or Nzyalipoovan, Rasthali or Morthaman, Karpuravalli or Pisang Awak, and Red banana that are dessert varieties grown in plains. Virupakshi and Sirumalai are hill banana varieties, while Monthan, and Batheesa are cooking types. Nendran is a dual purpose plantain mainly grown in Tamil Nadu and Kerala.

2.2 Production

In India, banana and plantains are cultivated in an area of 8.30 lakh hectares with a total production of 29.78 million tonnes. Banana occupies 13% of the total area under fruit crops and contributes 39.77% of total fruit production in the country. The productivity of banana in India is one of the highest in the world. The major bananas producing states are Tamil Nadu, Maharashtra, Gujarat, Andhra Pradesh, Karnataka, Bihar, Madhya Pradesh, Uttar Pradesh,

Table 1: Statewise area, Production and Productivity of banana in India

State	2008-09			2009-10			2010-11		
	Area (000' ha)	Production (000' MT)	Productivity (MT/ha)	Area (000' ha)	Production (000'MT	Productivity (MT/ha)	Area (000' ha)	Production (000'MT)	Productivity (MT/ha)
Tamil Nadu	124.4	6667.0	53.6	113.7	4980.9	43.8	125.4	8253.0	65.8
Maharashtra	80.0	4960.0	62.0	85.0	5200.0	61.17	82.0	4303.0	52.5
Gujarat	60.9	3571.6	58.7	61.9	3779.8	61.0	64.7	3978.0	61.5
Andhra Pradesh	80.1	2804.0	35.0	80.6	2819.6	35	79.3	2774.8	35.0
Karnataka	75.4	1918.8	25.4	104.4	2132.3	20.4	111.8	2281.6	20.4
Madhya Pradesh	28.8	1498.0	51.9	33.0	1459.8	44.2	38.1	1719.6	45.2
Bihar	31.3	1373.6	43.9	31.5	1435.3	45.62	31.9	1517.1	47.6
Uttar Pradesh		In others		30.4	1138.6	37.4	32.4	1346.1	41.5
West Bengal	39.8	954.1	23.9	41.0	982.2	23.9	42.0	1010.1	24.0
Assam	47.6	852.6	17.8	53.4	805.2	15.0	47.6	723.6	15.2
Others	140.2	1617.4	11.5	135.5	1735.8	12.8	175.3	1873.1	10.7
Total	708.8	26217.2	37.0	770.3	26469.5	34.3	830.5	29779.9	35.9

(*Source*: NHB Horticultural Database, 2011)

West Bengal, Assam, Odisha and Kerala. The statewise area, production and productivity of banana in India is presented in the following table.

Table 2: Trends in Area, Production and Productivity of banana in India during the last 20 years

Year	Area (000'ha)	Production ('000 MT)	Productivity (tonnes/ha)
1991-92	383.9	7790.0	20.3
2001-02	466.2	14209.9	30.5
2002-03	475.3	13304.4	28.0
2003-04	498.6	13856.6	27.8
2004-05	589.5	16744.5	28.4
2005-06	569.5	18887.8	33.2
2006-07	604.0	20998.0	34.8
2007-08	658.0	23823.0	36.2
2008-09	709.0	26217.0	37.0
2009-10	770.3	26469.5	34.4
2010-2011	830.0	29780.0	35.9

Table 3: Percentage increase in area, production and productivity of banana since 1991-92.

Year	% Increase in Area	% Increase in Production	% Increase in Productivity
2001-02	21.44 (21.44)	82.41 (82.41)	50.25 (50.25)
2002-03	23.81 (1.95)	70.78 (-6.37)	37.93 (-8.20)
2003-04	29.88 (4.90)	77.65 (4.15)	36.95 (-0.71)
2004-05	53.58 (18.23)	144.95 (20.84)	39.90 (2.16)
2005-06	48.35 (-3.51)	142.46 (12.80)	63.55 (16.90)
2006-07	57.33 (6.06)	169.55 (11.17)	71.43 (4.82)
2007-08	71.40 (8.94)	205.82 (13.45)	78.33 (4.02)
2008-09	84.68 (7.75)	236.55 (10.05)	82.27 (2.21)
2009-2010	100.65 (8.65)	239.98 (0.96)	69.46 (-7.03)
2010-2011	116.20 (7.75)	282.29 (12.51)	76.85 (4.36)
Average	60.73	165.24	60.69

There has been a consistent increase in area under cultivation of banana except during 2005-06. This expansion of area has been at an average rate of 60.73% during the last 20 years, while the production has been increasing at an average rate of 165.24% and the productivity at 60.69% during the same period. This has been possible due to several factors like availability of elite planting material, improved production practices, advanced plant protection measures and timely intervention of government in providing subsidies for inputs.

2.3 Distribution

Though there are several varieties of banana belonging to different genomic groups, each area has preference for certain varieties. The varietal distribution of banana is summarized in the following table.

Table 4: Varietal Distribution of banana in India

Sl.No.	State	Cultivars
1.	Andhra Pradesh	Grand Naine (AAA), Robusta (AAA), Amritapani (AAB –Silk group), Thella Chakkrakeli (AAA), Karpurachakkrakeli (Poovan, AAB), Monthan (AAB), Yenagu Bontha (AAB)
2.	Assam	Jahaji (AAA), Borjahaji (AAA), Honda (AAA), Chinia (AAB), Manohar (ABB), Kachkola (ABB), Chini Champa (AB), Bhimkol (BBB), Attikol (BBB), Jatikol.
3.	Bihar	Grand Naine (AAA), Alpan (AAB), Chinia (AAB), Chini Champa (AB), Malbhog (AAB), Kothia (ABB), Muthia (ABB) and Gauria (ABB).
4.	Gujarat	Grand Naine (AAA), Robusta (AAA), Dwarf Cavendish (AAA), Lacatan (AAA), Harichal (AAA), Gandevi Selection (AAA).
5.	Karnataka	Grand Naine (AAA), Dwarf Cavendish (AAA), Robusta (AAA), Elakkibale (AB), Poovan (AAB), Rasabale (AAB), Monthan (ABB), Hill banana (AAB)
6.	Kerala	Nendran (AAB), Palayankodan (Poovan, AAB), Poovan (Rasthali, AAB), Monthan (ABB), Nzhalipoovan (Ney Poovan, AB), Red Banana (AAA).
7.	Maharashtra	Basrai (AAA), Robusta (AAA), Bhusavali (AAA), Mahalakshmi (AAA), Grand Naine (AAA), Lal Velchi (AAB), Safed Velchi (Ney Poovan, AB), Ardhapuri (AAA), Hanuman (AAA) and Rajeli (AAB, Nendran).
8.	Tamil Nadu	Grand Naine (AAA), Robusta (AAA), Poovan (AAB), Rasthali (AAB), Nendran (AAB), Karpuravalli (ABB), Sakkai (ABB), Peyan (AA), Matti (AA).
9.	West Bengal	Champa (AAB), Morthaman (AAB, Rasthali), Amrit Sagar (AAB), Gaint Governor (AAA), Lacatan (AAA) and Monthan (ABB).
10.	Orissa	Champa (AAB), Morthaman or Amritpani (AAB, Rasthali), Monthan (ABB), Chinia (AAB), Amrit Sagar (AAB), Lacatan (AAA)

(*Source* : NHB Horticultural Database, 2011)

2.4 World Scenario

Though bananas / plantains are produced in 130 countries round the world, ten major producing countries contribute two-thirds to the total global production (Table-5). The three major Asian producers (India, China and Philippines) do not distinguish between bananas and plantains, while African and south American countries show these both as separate commodities. In 2011, India led the banana production, producing 20% of the world crop of 145 million metric tonnes. Uganda was the second largest producer with 8% of world share and 95% of that was reported as plantains. The details of the top producers of banana in the world are depicted in the following table.

Table 5: Top ten producers of the banana/plantains in the world (2011)

Country	Production (Millon metric tonnes)	Percentage of world production
India	29.7	20
Uganda	11.1	8
China	10.7	7
Philippines	9.2	6
Ecuador	8.0	6
Brazil	7.3	5
Indonesia	6.1	4
Colombia	5.1	4
Cameroon	4.8	3
Tanzania	3.9	3
Others	49.6	34

Only 12% of total world production is exported, and 88% is consumed domestically. In many parts of the world, banana/plantains are produced for household food needs or local trade as can be seen from the fact that three largest producers of banana in the world are not visible as significant exporters. Therefore, each banana producing country has their local varieties in demand. But in international trade, the Cavendish (AAA group with different variants) is the main group of banana exported.

Chapter – 3

Varieties and Their Characteristics

Edible bananas are classified into several main groups and subgroups. In the first place is the diploid *M. acuminata* group which originated in Malaya, Indonesia, the Philippines, Southern India, East Africa, Burma, Thailand, the West Indies, Colombia and Brazil. This group is characterized by dark-brown sheaths of pseudostem, and the leaves are nearly free from wax. The bunches are small and the fruits small, thin-skinned and sweet. Cultivars of this group are more important in New Guinea than any where else. Lady Finger which was popular in Queensland and New South Wales belonged to this group.

In second place, there was a group represented by the prominent and widely cultivated **'Gros Michel'** originally from Burma, Thailand, Malaya, Indonesia and Ceylon. It was introduced into Martinique early in the 19th Century by a French naval officer and, a few years later, was taken to Jamaica; from there it was carried to Fiji, Nicaragua, Hawaii and Australia, in that sequence. It is a large, tall plant bearing long bunches of large, yellow fruits, and it was formerly the leading commercial cultivar in Central Africa, Latin America and the Caribbean, but has been phased out because of its great susceptibility to Panama disease. It has given rise to several named sports or mutants (Mortan, 1987).

The Cavendish subgroup includes several important bananas which have traded all over the world. Important ones among them are

a) **'Dwarf Cavendish'**, was first known from China and widely cultivated, especially in the Canary Islands, East Africa and South Africa. The fruit is of medium size, of good quality and thin-skinned. It is susceptible to finger drop after ripening.

b) **'Giant Cavendish'**, (Syn. 'Mons Mari, 'Williams', 'Williams Hybrid', or 'Grand Naine') is of uncertain origin but it closely resembles the 'Gros Michel', and has replaced the 'Dwarf Cavendish' in Colombia, Australia, Martinique, in many Hawaiian plantations, and to some extent in Ecuador. The bunch is long and cylindrical, and the fruits are larger than those of the 'Dwarf Cavendish' and not as delicate as that.

c) **'Pisang masak hijau'**, or 'Bungulan', the triploid Cavendish clone of the Philippines, Indonesia and Malaya, is erroneously called 'Lacatan' in Jamaica where it replaced 'Gros Michel' because of its immunity to Panama disease.However, it is subject to Sigatoka leaf spot. The plant is tall and slender and prone to wind injury. It is no longer grown commercially in Jamaica and the Windward Islands. The fruits are commonly used as cooking bananas in Jamaican households. Simmonds declares this cultivar is not the true 'Lacatan' of the Philippines. He suggested that 'Pisang masak hijau' may have been the primary source of all the members of the Cavendish group.

d) **'Robusta'**, very similar to the so-called 'Lacatan', has largely replaced that cultivar in Jamaica and the Windward Islands and the 'Gros Michel' in Central America because it is shorter, thick-stemmed, less subject to wind. It is being grown commercially also in Brazil, eastern Australia, Samoa and Fiji. It is resistant to Panama disease but prone to Sigatoka.

Besides the Cavendish, Mysore and Silk group bananas are the next widely cultivated bananas in different parts of the world.

'Mysore', also known as 'Fillbasket' and 'Poovan', is the most important banana type of India. It is grown to a limited extent in Malaya, Thailand, Ceylon and Burma. It is thought to have been introduced into Dominice in 1900 but the only place where it is of any importance in the New World is Trinidad where it is cultivated as shade for cacao. It bears large, compact bunches of medium sized, plump, thin skinned, attractive, bright yellow fruits of sub-acid flavor.

'Silk', 'Silk Fig', or 'Apple' ('Manzana' in Spanish), is the most popular dessert banana of the tropics. It is widely distributed around the tropics and subtropics but never grown on a large scale. The plump bananas are 4 to 6 inches (10-15 cm) long, slightly curved; astringent when unripe but pleasantly sub-acid when fully ripe; and apple scented. If left on the bunch until fully developed, the thin skin splits lengthwise and breaks at the stem end causing the fruit to fall.

Among the plantains, there are many forms, some with pink, red or dark-brown leaf sheaths, and also some having colored midribs or splotches on leaves or fruits. The plants are usually large, vigorous and resistant to Panama disease and Sigatoka but attacked by borers. Major subgroups are known as 'French plantain' and 'Horn plantain'. The usually large, angled fruits are borne in few hands. All are important sources of food in southern India, East Africa, tropical America and the West Indies.

Though there are 942 accessions of bananas with distinct features maintained at National Research Centre for Banana, Trichy in Tamil Nadu. Many of them are same varieties with different synonyms grown or found in different parts of the world. Based on their genetic composition these have been reduced to nearly 250 varieties / land races with distinct features. Out of these, less than fifty are commercially cultivated in different parts of India. The specific characteristics of these genotypes are detailed below along with their synonyms (Uma *et al.*, 2005). Though successful breeding in banana is very difficult, focused effort by scientists of FHIA Honduras and in SAUs of India has resulted in development of hybrids which are also detailed below.

Dwarf Cavendish (AAA): This has several synonyms like DC, Basrai, Bhusawal, Jahaji, Kabuli, Pacha vazhai, Mauritius, Morris, Kuzhi vazhai, Sindhurni and Singapuri. This belongs to AAA genomic group and was widely cultivated in Gujarat, Maharashtra, Madhya Pradesh, Bihar and Karnataka. Now in many places, it is replaced by Grand Naine. The plant is of short stature growing to a height of 1.5 meters above the ground level. The pseudostem is characterized by brown-black large blotches of conspicuous size spread all over, with a girth of 75-80 cm. The leaves are clustered at the crown with short internodes. The petiole is winged with a widely open canal without clasping the pseudostem. The inflorescence is pendulus with female phase having 8-12 hands at regular interval. The fruits are borne clusters of 16-24 with a short pedicel and blunt tip. Bunch is cylindrical and fruits are yellowish green after ripening. It develops brown spots at over-ripe stage in sub-tropical conditions.

Grand Naine: This is a tall mutant of Dwarf Cavendish. It bears bunches weighing 25-30 kg with well spaced hands and uniform sized long fingers. Fruit quality is almost like Dwarf Cavendish. It is early maturing variety and normally comes to harvest 11 months after planting.

Robusta (AAA): This is a semi-tall sport of Pacha Vazhai or DC and is also known as Bombay Green and Harichal in Maharashtra, Pedda Pacha Arati in Andhra Pradesh and Borjahaji in eastern India. The plant is medium tall with 2-2.5 meters height and the psuedostem is blotched. The petiolar wings do not clasp the peudostem and the inflorescence is pendulous. The bunch is cylindrical with 10-12 hands and weighing 25-30 kg. Each hand has 16-24 fingers arranged in two rows. Each finger has a short pedicel and blunt tip. After ripening, the fruits turn to bright yellow colour. The fruits have a long green life and short yellow life. At over-ripe stage fruits drop from pedicel.

Red Banana (AAA): Red banana is also known as Lal Kela, Chendkadali, Chevvazhai, Yerra arati, Chandra bale and Kembale in different states of southern India. This is commercially cultivated in Tamil Nadu and Kerala states and is highly priced variety. It is believed to have many medicinal properties. The plant is tall, growing to a height of 2.75 -3 meters, with a purple red pigmented pseudostem, leaf pedicel and fruit peel. The pulp is sweet, orangish yellow in colour and has a pleasant aroma. It is a shy yielder but has a potential to yield even 30-40 kg per bunch. Normally it has 5-8 hands with 12-14 fingers in each hand. The fruits are slightly curved, with blunt tip and thick purple skin.

Venkadali (AAA): This is a bud-sport of red banana with a normal green pseudostem. The fruit skin is green unlike red banana and turn yellow when ripe. The texture and taste of fruit is similar to red banana.

Thella Chakkarakeli (AAA): The synonyms of this variety are Chakkarakeli, Honda, Rajabale, and Raja Vazhai. It is commercially grown in Andhra Pradesh, while in Karnataka, Tamil Nadu and North Eastern India it is backyard vareity. This is a short statured variety growing to a height of 2.25-2.4 meters with greenish yellow pseudostem. The petiolar wings are broad with a conspicuous reddish pink margin. It puts forth a bunch having 4-6 hands each with 12-14 stout fingers which have a blunt tip. The fruits are greenish yellow in colour which turn to golden yellow when ripe. The pulp is sweet and juicy with pleasant aroma.

Matti (AA): This is a diploid variety belonging to *Musa accuminata*, and largely grown in Kanyakumari and Tirunelveli districts of Tamil Nadu. It is a medium tall variety growing to a height of 2-2.5 meters with yellowish green pseudostem having brown blotches. Normally a bunch having 12-15 hands weighs 18-20 kg. Each hand bears about 14-20 fruits. The fruits upon ripening have sweet and creamy coloured pulp with characteristic aroma.

Ney Poovan (AB): Ney Poovan is a unique seedless diploid also known as Nzhalipoovan, Elakki Bale, Safed Velchi and Chini Champa in different parts of the country. Kunnan and Nattu Poovan are two others belonging to this sub-group. This is a slender medium tall stature plant with thin stem having uneven spread of brown-black blotches. It produces 10-12 hands each having 14-16 fingers closely packed. It is variety having good keeping quality and almost non-shattering even at over-ripe stage. The fruit is sweet with smooth textured pulp.

Rasthali (Silk AAB): Rasthali is the choicest variety grown in Tamil Nadu, Andhra Pradesh, Kerala, Karnataka and Bihar. Rasabale, Malbhog, Amrithapani, Rasakeli, Morthman and Poovan are the other synonyms of this variety. It is a medium tall statured plant with thick pseudostem which is yellowish green in colour. The bunch bear 12-14 hands each with 12-16 fingers. The fruits are golden yellow after ripening. The pulp is very tasty with good sugar acid blend, pleasant aroma and firm texture. The over-sized fruit sometimes have lumps. The fingers detach and shatter from the crown at over-ripe stage.

Poovan (Mysore AAB): Poovan is grown all over the country and is a leading cultivar in Tamil Nadu, Kerala, Andhra Pradesh and Karnataka as well as Northeastern Region. It has several synonyms like Champa, Alpan, Dora vazhai, Karpura Chakkrakeli, Mysore, Palayankodan, etc. It is a medium statured plant with characteristic pink pigmentation on pseudostem. The bunch is cylindrical in shape with closely packed, short, stout and beaked fruits. A bunch weighs 18-25 kg with 12-15 hands, each having 12-16 fruits. On ripening the fruits turn golden yellow and the pulp is sour-sweet in taste. It has a short yellow life and, susceptible to fast spoilage after ripening.

Nendran (Plantain AAB): Nendran belongs to the French plantain type and is widely cultivated in Kerala and Tamil Nadu. It is believed to have originated in Southern India and spread from there to Africa. The plant is medium statured with slender pseudostem, which is yellowish green in colour with pink tinge. It is a short duration variety which comes to harvest 11 months after planting. The bunches bear 4-6 hands, each with 8-10 long fruits. It is a low yielder with an average bunch weight of 12 kg. The fruits are quite starchy even after ripening. The skin turns black upon full ripening but fingers do not shatter even at over-ripe stage. More than 50% of the total production this variety goes for processing as chips, while rest is consumed after steaming or boiling. In Kerala state, it is a breakfast fruit. There are several ecotypes in this variety *viz.*, Moongli, Zanzibar, Manjeri Nendran, Myndoli, Attu Nendran, Chengali Kodan and Changanacheri Nendran. The pulp of this variety turn golden yellow upon ripening and is richest in carotenoids among the bananas.

Hill Banana (Pome, AAB): Varieties like Virupakshi, Sirumalai, Malavazhai, Vannan, Marabale and Ladan belong to this group. Its wide cultivation is limited to Palani and Sevroy hills of Tamil Nadu. It is the only variety under perennial cultivation system in India. The plants are usually tall with greenish yellow pseudostem having dark blotches spread all over. The bunch is small and takes slightly longer duration to mature. The fruits turn yellowish green upon ripening with cream coloured pulp. It is relished for its distinct flavour. It is a preferred variety for making panchamrutham (Prasad), in the temple of Palani.

Monthan (ABB): Monthan is primarily grown for culinary purposes. Both the raw green fruits and true stem are used for cooking various kinds of dishes. Sometimes it is also used for making chips. It is widely cultivated in all the banana growing areas and known by different names like Bontha, Nalla bontha, Karibale, Bainsa, Kashkal, Bankel, Kachkola, Jatikol, *etc.* It has several mutants like Thella bontha, Sambal monthan, Nalla bontha batheesa, Sambarani monthan and Booditha montha batheesa. The plant is tall and robust with green, shiny pseudostem. It is short duration crop which comes to harvest within 12 months. The bunch bears 12-14 hands, each with 5-6 stout and angular fruits. The average bunch weight is 18 kg.

Karpuravalli (ABB): Karpuravalli also known as, Kanthali in Bihar, Chara Poovan, Bhat Manohar, Pisang Awak, Pey Kunnan and Kostha Bontha in other states is a popular variety grown both for dessert as well as cooking purposes. The plant is tall growing to a height of 3 meters with pseudostem having pink streaks. It is long duration variety taking about 15-16 months to mature after planting. Each bunch weighs about 25-30 kg with 12-15 hands. Each hand consists of 12-16 fruits which are ash coated, medium in size and having a conspicuous beak. The pulp is white in color even after ripening, very sweet in taste with a TSS of 28-32°Brix . It has very good keeping quality, non-shattering, firm and edible even after skin turns black. It is highly suitable for making banana fig, wine and juice.

Chakia (ABB): It is a dual purpose variety popularly grown in Bihar, Tamil Nadu and Kerala. It is also known by its synonyms like Sakkai, Gauria and Muthia (Northeast region). It is a tall variety with sturdy pseudostem having blotches. It takes 12-14 months to mature after planting. Each bunch weighs about 15-18 kg with 5-8 hands. Each hand has 12-14 fingers which are short and stout without knob. The fruits turn dirty yellow on ripening and is sweet-sour in taste.

Bhimkol (BBB): This is triploid *Musa bulbiciana* variety naturally found in Northeastern India, known for its medicinal value and popular as baby food. The plant is very tall and robust growing to a height of 4.5 meters. The pseudostem is green in colour without any blotch or pigmentation. It take 16-17 months to produce a mature bunch after planting. Each bunch weighs about 22-25 kg with 6-8 hands and each hand having 8-12 fruits. The fruits are short and stout with prominent ridges. The fruits have many seeds enveloped with a thin layer of pulp. The pulp as well as other parts of the plant are known for various medicinal values.

Hybrids: Development of hybrids in banana is very difficult due to its recalcitrant nature. Though banana breeding programme in India was started in 1949 no significant results could be achieved. Of late the systematic research in banana breeding was undertaken at TNAU, KAU and NRCB which resulted in development of some hybrids like H-1 (Agniswar x Pisang linin); H-2 (Vannan x Pisang linin); Co-1(3 way cross between Kallar Ladan x M.bulbisiana clone Sawai x Kadali); H-59 (Matti x Anaikomban); H-65 (Matti x Anaikomban) ; H-109 (Matti x Tongat). Besides these, global hybrids developed by FHIA, Honduras, viz., FHIA-01 (Gold Finger-AAAB); FHIA-02 (Williams x SH 3393), FHIA-03 (Saba hybrid) (SH-3386 x SH-3320), FHIA-17 (Highgate x SH-3362) and FHIA-21 (AVP-67 x SH-3142) have several desirable traits like resistance to black sigatoka, fusarium wilt, resistance to burrowing nematode, drought tolerance and salt tolerance. During their evaluation world wide, these have proved their superiority in Australia, Brazil, Cuba and has promises for many other countries. However, though these have shown great promise, none of these hybrids could impress the banana growers and replace the existing varieties in India. Brief descriptions of some of these hybrids are given below.

BRS-1 (Agniswar x Pisang Lilin): It is a promising hybrid owing to short cropping cycle, immunity to leaf spot, fusarium wilt and burrowing nematode (*Radopholus similis*). It is a medium tall plant, supporting 14-16 kg bunch without propping. Elongated fruits turn attractive golden yellow upon ripening. It is slightly acidic which vanishes upon full ripening with high sugar content. BRS-1 has a remarkable faster rate of ratooning ability completing four crop cycles in three years. Multilocational trials have shown its acceptability among growers and consumers especially in Andhra Pradesh. Immunity to leaf spot diseases, one of its unique traits, has made this hybrid highly appreciable among growers.

BRS-2: A hybrid of Vannan x Pisang Lilin developed at Banana Research Station (KAU) Kannara is a medium-statured plant growing up to 2-2.5 m. Crop cycle is short, only 11- 12 months. The bunch weight ranges from 15-20 kg with short, stout, dark green Poovan like fruits, which are arranged very compactly. Fruits are slightly acidic with pleasant sweet-sour aroma. Its immunity to leaf spot diseases and nematodes makes it a suitable for subsistence cultivation.

Co-1: This is one of the AB hybrids from the cross of Kallar Ladan (AAB) x *M. balbisiana* clone Sawai was used to cross with cv. Kadali (AA) to develop an AAB Pome hybrid. This was later released for commercial cultivation as Co-1 banana. This belongs to Pome group and closely resembles Virupakshi (AAB-Pome), another popular hill banana in Tamil Nadu. It retains the typical acid/apple flavour of Virupakshi even when grown in plains contrary to Virupakshi, which develops aroma only when grown at higher altitudes.

Screening of some of the synthetic banana hybrids and their parents against fusarium wilt of banana (*Fusarium oxysporum* f.sp. *cubense*, race 1) under artificial conditions showed that the hybrids namely, H.65, H.103, H.109 and H.201 and parents viz. Anaikomban, Pisang Lilin and Robusta exhibited distinct resistant reactions. Some of the new progenies evolved by inter diploid crosses involving diploids synthesised earlier were also evaluated. Preliminary investigations indicate that the hybrids H. 201 x Anaikomban (tetraploid) and H-59 x Ambalakadali to be highly resistant to sigatoka diseases. The newly evolved synthetic hybrids *viz*., H-203, H-205, H-205, H-207 and H-213, H-234 and H-237 also appear to have higher tolerance to nematodes. The hybrids H-203 and H-218 also exhibit higher resistance to sigatoka leaf diseases.

FHIA-01 (Gold Finger): It is pome hybrid (SH-3142 x Dwarf Prata) from Honduras with AAAB genome. Fruits are sub-acid, have good shelf life and the bunch weighs 20-25 kg. It is resistant to black sigatoka, fusarium wilt and burrowing nematode.

FHIA-02: It is tetraploid (AAAA) developed at Honduras by crossing Williams x SH-3393. This hybrid is similar to Cavendish and resistant to sigatoka but susceptible to fusarium wilt. The plant grows tall with 2.5-3.0 m height and comes to harvest in 12-13 months. The bunch weight ranges from 15-20 kg and the fruits have short postharvest green life.

FHIA-03: It is cooking type hybrid having a tetraploid constitution (AABB). It was developed from (SH-3386 x SH-3320) in Honduras. The plant is vigorous type growing to a height of 2.5 to 3.0 m. It is dual purpose hydrid, but more suitable for cooking. It is suitable for backyard cultivation and is resistant to sigatoka leaf spot.

Chapter – 4

Trade –International and Domestic

4.1 Importance of Banana in International Trade

Bananas are among the most widely consumed and globally traded foods in the world after rice, wheat and maize. It provides food security and accessibility in dozens of countries in the developing world. While in some developing countries it is a part of staple diet, for others it supplements the nutrition. Most producers are small-scale farmers and produce either for home consumption or local markets. Because bananas and plantains produce fruit year-round, they provide an extremely valuable food source during the *hunger season*. Bananas and plantains are, therefore, critical to global food security. At the same time bananas are the main fruit in international trade and the most popular one in the world. In terms of volume they are the largest exported fruit, while they rank second after citrus fruit in terms of value. Banana is a very delicate commodity on economic, social, environmental and political grounds.

Total world exports at around 18.3 million metric tonnes amounted to only 12% of total world production; two thirds of the exports were generated by only five countries *viz*., India, China, Philippines, Costa Rica and Gautemala. The top three producing countries do not contribute much to export trade while the later two (Costa Rica and Guatemala) along with Ecuador and Colombia contributes 66% to total world exports. Among the top ten producers only the

Philippines has a consistent position both as a producer and exporter (9%) from Asia in Table 6.

Table 6: Top ten exporters of the banana/plantains in the world (2011)

Country	Exports (Millon metric tonnes)	Percentage of world exports
Ecuador	5.2	29
Costa Rica	1.8	10
Colombia	1.8	10
Philippines	1.6	9
Guatemala	1.5	8
Others	6.0	34
Total	17.9	100

According to the Food and Agriculture Organization of the United Nations (FAO) Statistics, world total exports of banana accounted for 18.3 million tonnes in 2009. In 2009, the five leading banana-exporting countries were Ecuador, Colombia, the Philippines, Costa Rica, and Guatemala. Together, they accounted for approximately 84 percent of the global banana exports in 2009. Ecuador is the main supplier of bananas in the world market, with exports of 5.7 million metric tonnes in 2009, which is equivalent to 38 percent of the total volume of bananas traded that year. It was followed by Colombia, with a market share of 13.2 percent, the Philippines (11.7%), Costa Rica (11.5%), and Guatemala (9.9%). By region, Latin America is the world's top supplier of bananas, supplying 83.2 percent of the gross fresh bananas exported in 2009, followed by the Far East at 12.8 percent, Africa at 3.4 percent, and the Caribbean at 0.4 percent.

4.2 Major Exporting Countries

While examining the history of international banana trade, it can be seen that it has tripled between the 1970s and 2009. But world trade is still characterised by a high concentration of key players: five countries, four of which are in Latin America - Ecuador, Colombia, Costa Rica, Guatemala and one Asian - the Philippines. This trend has continued into 2010 & 2011 with the four Latin American countries exporting 83% of the total with 11.6 million metric tonnes out of a worldwide exports in 2010 (13.9 mMt). The banana industry is a very important source of income, employment and export earnings for these major banana exporting countries. According to UN COMTRADE database, world banana exports are valued a total of US$ 8.2 billion in 2011 (+3.6% as compared to 2010), making them clearly a vital source of earnings to many countries.

According to FAO Statistics during 2011, the total world export of bananas and plantains was 14.939 million tonnes. Out of this 12.193 million tonnes (81.68%) came from Latin America & Caribbean countries, 2.165 million tonnes (14.49%) from Far East mainly the Philippines- 2.046 million tonnes (13.70%), China, Malaysia, Pakistan, Thailand, Vietnam, India, Indonesia, and the rest (0.580 million tonnes) from Africa and Oceania Islands (3.88%).

4.3 Major Importing Countries

Global imports of fresh bananas have followed a long-term upward trend, reaching 18.3 million metric tonnes in 2009. The major importers are US, EU and Japan. In 2009, the United States, the European Union, and Japan accounted for about 56 percent of the total world's fresh banana imports. Over the period of 2000 to 2009 the imports of bananas to the United States and Japan remained relatively flat , compared to the European Union where it increased by 15.6 percent, from 3.9 million metric tonnes in 2000 to 4.5 million metric tonnes in 2009. The EU banana market is controlled by custom tariffs and market intervention. For example, the European Union imposes tariffs on bananas from Latin America while allowing bananas imported from the African, Caribbean, and Pacific (ACP) countries to enter duty-free considering their pre-colonial status. This caused the widely known banana trade war, which resulted in the European Union cutting duties on bananas from Latin America gradually to 144 Euros per metric tonne by 2016 from the current 176 Euros per metric tonne (Evans and Ballen, 2012).

4.4 World Market Structure for Fresh Bananas

A close look at the world's fresh banana market shows that it is characterized by an oligopolistic market, with only a few multinational companies (MNCs) engaged in the purchase, transport, and marketing of the fruit. Typically, these companies are integrated vertically with their own or contract plantations, own sea transport and ripening facilities, and have their own distribution networks, which gives them considerable economies of scale and market power in selling bananas. The world's three largest producers and marketers of fresh bananas are Dole Food Company, Chiquita Brands International, and Del Monte Fresh Produce. Each of these companies owns banana plantations in most of the banana producing regions in the world. The export data of these countries shows that the estimated market share of the six major banana producers/ marketers in 2009 was 80 percent as follows:

Chiquita, Del Monte, Dole, Fyffes and Bonita Banana control around 80% of the sales on the banana import market worldwide. The two largest are U.S. based corporations, Dole Food Company and Chiquita Brands International, control about 25% each of the market globally. The three other major corporations that control the banana trade are Fresh Del Monte Produce (Chilean-based IAT Group) with a 15% share, Bonita Banana (Ecuadorian-based Exportadora Bananera Noboa, part of the conglomerate Gropu Noboa) with a 9% share, and Fyffes (Irish-based) with a 7% share.

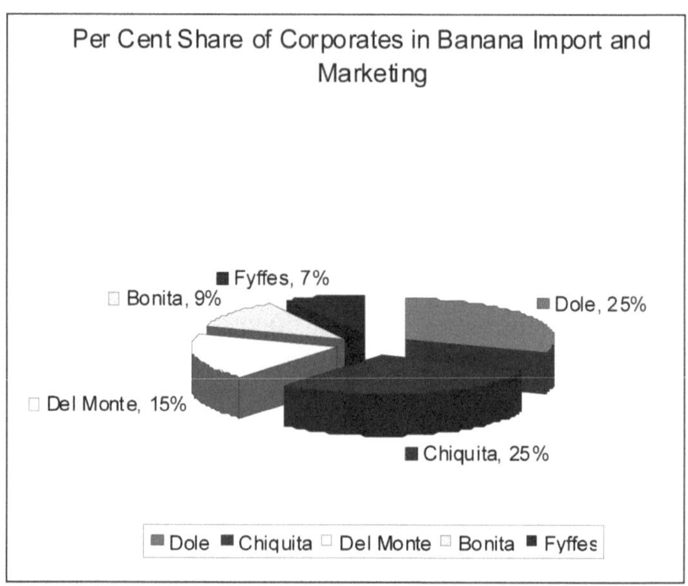

Though India is the largest producer of banana with a total production of 29.7 million tonnes, (20% of the global production), its contribution to global banana exports is only 0.3%. The Indian banana exports goes mainly to UAE, Saudi Arabia, Iran, Kuwait, Bahrain, Qatar, Oman, Nepal and Maldives, bringing in approximately Rs.1000 million (~ US$ 20 million). Like many other banana producing countries, its cultivation is still one of the main stakes for livelihood for small and marginal farmers in India.

4.5 Domestic Trade

Almost 99% of banana produced in India is consumed domestically. The quantity of internationally traded bananas is about 18.3 million tonnes to which India contributes only 54,304 tonnes (in 2011), forming a negligible 0.30% (NHB, 2011). Though the banana are grown in many states, the supplies are available year round only in some states like Maharashtra, Tamil Nadu, Andhra Pradesh and Goa in mainland and in Arunachal Pradesh, Assam, Nagaland, Tripura in northeastern region. In rest of the areas the peak season varies, as there is difference in the time of planting. In Bihar, UP, Gujarat,

Jharkhand, Madhya Pradesh and Uttarakhand of northern region, and in Karnataka and Kerala of southern region the peak harvest season is between July to November. In eastern regions like Orissa, West Bengal and Sikkim the main harvest season is during August to October. In the Andaman and Nicobar Islands the peak season is during October to March. The location specific varieties are mostly sold in the local markets, while the Cavendish bananas are transported over longer distance to major markets. The major arrival markets are metropolitan cities like Chennai, Bangalore, Hyderabad, Mumbai, Delhi and Kolkata. From these markets the produce moves to both interiors areas in the same state as well as across the states.

Among the domestic markets, Bengaluru, Chennai, Delhi and Mumbai are the major destinations where large quantities of bananas arrive during the year. Chennai receives the maximum produce among the four metros, with around 1.57 lakh tonnes annually. Maximum banana trade (90,640 tonnes) in Chennai is seen during August to December months. In Bengaluru the peak arrivals are in May, June, July and August, during which 38-40% (44, 600 tonnes) of total annual arrivals (1.16 lakh tonnes) are witnessed. Delhi market receives majority of its supplies (60,000 tonnes) during July to October. Mumbai market gets maximum bananas during July to November. Other important markets are Hyderabad, Kolkata, Jaipur, Guwahati, Patna, Surat and Srinagar. The average whole sale market prices of banana range from Rs.723 to Rs.2046 per quintal. The prices at any point of time is proportional to arrivals of the fruits in smaller markets, while in major markets like Chennai and Bengaluru the prices are higher during the peak arrival season (August to November in Chennai and May to August in Bengaluru). Annual average prices of bananas are highest in Pune (Rs.2046/Q) followed by Dehradun (Rs.1983/Q), Guwahati (Rs.1980/Q), Hyderabad (Rs.1956/Q) and Srinagar (Rs.1930/Q).

4.6 Banana Trade Wars: A Glimpse

The international trade in banana had become a matter of global dispute between countries, as the economy of some countries were dependent on banana trade. About 95 per cent of U.S. households buy bananas every week making it a significant part in regular budgets of American families. In US, spending on bananas is only slightly lower than the spending on gasoline. Bananas are the best selling fruit in the UK: the Britons eat over 3 billion bananas a year. Because of this fact banana became a matter of serious business.

The banana has an extensive trade history starting with firms such as Fyffes and the United Fruit Company (now Chiquita) at the end of the 19th century. For much of the 20th century, bananas and coffee dominated the export economies of Central America. In the 1930s, bananas and coffee made up as much as 75% of the region's exports. The United Fruit Company based its business almost entirely on the banana trade, because the coffee trade proved too difficult to control. The term 'banana republic' has been applied to most countries in Central America, though in strict economic perspective only Costa Rica, Honduras, and Panama had economies dominated by the banana trade. The European Union has traditionally imported many of their bananas from former European Caribbean colonies, paying guaranteed prices above global market rates. This was a kind of preferential treatment meted to Caribbean countries by Europe.

The U.S. corporations, working mostly in Latin America, dominate the international market. Latin America and the Caribbean region provide about 80 percent of bananas for the world market. Another major exporter is Africa's Uganda. The struggle for the European market worsened in 1991, when representatives of the powerful banana republic of Costa Rica accused European countries of protectionism that grant preferences to African exporters of bananas.

In developed countries like US & UK the farmers' price realization from exports was lowest due to discounted price being paid by the big grocery retail giants for bulk buying. Price competition among grocers has reduced their margins, leading to lower prices for growers. In US and Europe the super markets are now the only players in the banana chain to consistently make profits from bananas. They have dramatically increased their economic power in the banana chain in the last decade and more. Bananas are the single most profitable item in British supermarkets, accounting for 1% of their total sales. In the USA, it is estimated that bananas represent 2% of the total turnover of North American grocery retailers. UK have a large concentration of supermarkets controlling most of the sector, leading to the usual economic and social problems that concentration brings. It is reported that supermarkets are not neutral actors in banana trade and that they are largely responsible for their aggressive price cutting (by often making suppliers reduce their prices rather than reducing their own margins, though sometimes their own margins are already low). This resulted in low wages for workers in banana plantations of exporting countries. Some workers work very long hours, sometimes exposed to hazardous chemicals, some which are often banned in nations such as the US and Europe. Child labor, gender discrimination and other problems accompanied this drive for low consumer prices.

Caribbean bananas are grown on small, family-run farms. In Central America the farms are large and owned by wealthy farmers and companies. The producers in Latin America, with the United States behind them, accused the EU of providing tax benefits to banana producers from Africa, the Caribbean and the

Pacific. Indeed, they were actually preferences given to European colonies under the so-called Lomé Conventions. In 1993 a system of quotas was established; tariffs for suppliers from different regions were different. In 1994, several countries in Latin America signed a separate agreement with the EU to allocate fixed delivery quotas for them: 23.4 percent - for Costa Rica, 21 percent - for Colombia, 3 percent - to Nicaragua and 2 percent - for Venezuela.

In 1997 World Trade Organization (WTO) decision (pressured by the US, backed by companies like Chiquita), meant that these local producers will have to compete on a "level playing field" with giant multinationals and Latin-American "dollar" bananas. According to the WTO ruling, there shouldn't be discrimination based on where the food is produced or even how it is created (which also means, for example, that a country cannot easily say no to genetically modified food and the general public doesn't have to be informed). This has affected the way the European Union (EU) had it's trade agreements with the African, Caribbean and Pacific countries (ACP countries) and the Lomé; Convention which is based on cooperation and partnership between the EU and 71 ACP member states. The convention was the world's largest trade and aid pact and due for expiry in 2000. It had to be replaced by a system designed to open up the markets of the poor nations to the industrialized nations of Europe. In 1998, UK already voted to abolish the Caribbean's guaranteed access to European markets. In its place a proposal for the EU to issue import licenses to ships arriving with bananas from ACP countries on a 'first come, first served' basis was made. The ACP nations felt that they were not be ready to enter a global market place with free trade in the way that the WTO prescribes. By 2005, these arrangements were in the process of being withdrawn under pressure from other major trading powers, principally the United States..

The European Union and ten Latin American countries have signed an agreement at the headquarters of the World Trade Organization (WTO) which was expected to put an end to the "banana war" that began in 1991, when largest transnational corporations under the cover of the governments of Latin American countries accused Europe of protectionism. It is said that the war with the assistance of the WTO was conducted between the U.S.-based Chiquita Brands and Dole Fruit on the one hand, and on the other hand - Geest, Fyffs (UK), Bargosa (Spain) and smaller companies, whose interests were defended by the EU.

The European Union gave up in 2008, and later, the European Parliament approved the agreement. It was agreed that by 2017, the duties on bananas from Latin America are to be cut by 35 percent - from 176 to 114 euros per ton. The benefit goes to such Latin American countries as Brazil, Venezuela, Guatemala, Honduras, Colombia, Costa Rica, Mexico, Nicaragua, Panama,

Peru and Ecuador. Since 2009, the agreement has been ratified by all EU countries and all the procedures were finally completed in 2012, bring and end to the 20 years long banana trade war. Latin American bananas currently dominate the European market, despite the duties.

4.7 Fair Trade Bananas

In international banana industry, production, profits, and market access are highly concentrated, where just five corporations control (i.e. Dole, Chiquita, Del Monte, Bonita and Fyffes) who control 80% of sales in import trade through out the world. Only about 20 % of the prices paid by consumers in supermarkets reach exporting country. The lives of banana plantation workers have been pathetic due to low wages, over work (12-14 hours per day) and poor welfare measures. Most of the plantations of these corporations are based in Latin American countries, where the fruits are produced at the cheapest price possible, often voilating the labour rights and environmental protection regulations. The typical banana plantation in Central America uses up to 70 kilograms of pesticides per hectare per year – over 10 times more than is used in the production of other crops in industrialized countries. These chemical sprays may have a serious impact on the health of workers and people living in the area, as well as the surrounding wildlife. As a result of the steady decrease in banana prices over the past decades, the daily life of many plantation workers and small farmers in producer countries is deteriorating.

Fair Trade foundation is one which protects the interests of small banana growers and plantations. Since the first launch in the Netherlands in 1996, fair trade banana sales have been multiplied by 4 and fair trade bananas are now available in the Netherlands, Denmark, Germany, Sweden, Belgium, Luxembourg, Austria, the UK and Switzerland. They come from 12 producer groups in Ecuador, Colombia, Costa Rica, Ghana, the Dominican Republic and the Windward Islands, representing some 5,000 disadvantaged farmers or workers and their families.

Fair trade can help secure market access for the most vulnerable farmers. Fair Trade Marketing initiatives exist in Austria, Belgium, Denmark, France, Germany, Luxembourg, Netherlands, United Kingdom and Ireland. They seek to change unfair international trading structures and improve the social, environmental and economic conditions of disadvantaged producers by giving banana producers and workers direct access to a market, guaranteeing better trading and working conditions and thus providing them with the tools that permit them to control their own development, and to invest in environmentally friendly production methods.

Chapter – 5

Consumption Pattern of Bananas and Plantains

Most banana producers in Africa, Latin America and Asia are small-scale farmers who grow bananas either for home consumption or local markets. Because bananas and plantains produce fruit year-round, they provide an extremely valuable food source during the *hunger season* (when the food from one annual/semi-annual harvest has been consumed, and the next is still to come). Bananas and plantains are therefore critical to global food security. However, the export trade of banana depends mainly on commercial plantations in Latin America, East Asia and Africa. These plantations are owned by multinational corporations.

World consumption per capita of bananas and plantains period of 1985 to 2002 grew at a rate of 1 percent per annum, and signals a shift from the stagnant consumption of the previous 15 years (1970-1984). The increase was due to higher desert banana consumption in the developed world, where the rate of growth was 4.2 percent per annum throughout the period. By contrast, consumption in the developing world remained stagnant and in the newcomers to the world economy (for example Eastern Europe and China) it increased moderately. World consumption in 1998 - 2000 stood at an average 15.3 kg per capita; developed countries at some 13 kg per capita and developing countries at 21 kg (almost entirely domestically produced). The economies in transition consume small quantities of bananas but these are expected to increase as their economies grow stronger. In the period 1998-2000 the Russian Federation consumed 3 kg and China 3.5 kg per capita.

Still banana is the cheapest, plentiful and most nourishing of all fruits. It contains nearly all the essential nutrients including minerals and vitamins and has several medicinal properties. Banana is a rich source of energy. About 24 bananas each weighing around 100gm would provide the energy requirement (2400cal/ day) of a sedentary man. Besides as a fresh fruit it is also used in beverages, ice-creams, confectionary and baby foods. Culinary bananas are also used for making various kinds of traditional foods in various parts of the world.

Plantain and bananas (*Musa* spp.) serve as important food crops in much of Africa. Bananas are a major part of the staple diet, complementing other sources of food like beans and cassava. Banana and plantains together provide over 25% of the carbohydrate needs and 10% of the daily calorie requirements for over 70 million people on the African continent (Luntala *et al*, 2000 ; Harvestplus, 2007). Apart from this, some cultivars have high pro-vitamin A carotenoids (pVACs) levels and are capable of providing up to half of the total human daily vitamin A requirement in a single fruit (Davey *et al*, 2006, 2009a and 2009b). PVACs are much higher in plantains (upto 200-folds) than cooking or commercial dessert types (Harvestplus, 2007). In Uganda, bananas are estimated to provide 30% of calories, 10% of protein and 5% of fats for the entire population.

The bananas are consumed in different ways like dessert, cooked, baked, fried, dried, ground, fermented, etc. Cooking of banana fruits range from simple boiling to fermentation, drying and grinding to make flour; The fruit can also be eaten as a dessert, or processed into juice, beer and wine (Ndungo *et al.*, 2008). Other cooking methods practiced included deep frying the plantains in local palm oil. Boiling of the banana fruit is a common practice in Uganda, Cameroon, Burundi and West Indies. In Uganda, the fruit is boiled with beans and ghee then mixed with pepper, salt and onions to make a dish called 'akatogo'. In Cameroon, the green banana is boiled and served in a sauce of palm oil with fish, cooked meat, green beans and seasoning, whereas in the West Indies boiled green banana is served with salted fish or meat (FAO, 1990).

Apart from boiling the plantain varieties are also dried and ground into flour and mostly used for making porridge for children less than 5 years age in East Africa. The porridge made from plantain flour is especially useful for children with diarrhea. The banana flour could be mixed with cassava flour and not only used for making baby porridge but also boiled in water to make a thick paste locally called 'ugali.' The 'ugali' could be consumed with any relish, the most common one being, boiled cassava leaves, locally called 'sombe'. *Musa* flour especially from plantain banana is the only local *Musa* product that seemed to have a long shelf life. The same is true of practices in Gabon and Cameroon, where the mature green form of banana is dried and stored and may be used for cooking after grinding into flour, the difference is that in these two countries (Gabon and Cameroon) this product was mainly used as a famine reserve, not an everyday meal (FAO, 1990).

Steaming and roasting are very popular banana cooking methods in Uganda. Some banana cultivars are mostly steamed in banana leaves and either pounded or eaten whole (a dish called 'matoke'), while the plantains are roasted and sold as street food. However, these methods are not popular in Congo.

In India 55-60% of total bananas produced are dessert bananas and the rest are cooking bananas. Plantains (French plantain types) cultivation is mostly restricted to Kerala and Tamil Nadu. The dessert bananas (cream or white fleshed) are eaten as dessert after ripening. Unlike Africa, the cooking bananas in India are consumed as a vegetable after boiling and addition of spices and condiments and seasonings. Here it is not a major source of carbohydrates but is only supplementary. However, the ripe bananas considerably adds to caloric intake. The Nendran banana (plantain type) is consumed as a breakfast food after boiling in Kerala. More than 50% of plantains produced in India are used for making chips, which is a deep fried product with salt and pepper.

Chapter – 6

Nutritive, Medicinal and Therapeutic Values

Consumption of fruits and vegetables has been associated with a reduced risk of chronic diseases such as cardiovascular diseases and cancer. Phytochemicals, especially phenolics, in fruits and vegetables are suggested to be the major bioactive compounds responsible for health benefits. Banana fruit is one of the high calorie tropical fruit which provides 90 calories of energy per 100 g of fruit. Besides, it contains good amounts of health benefiting anti-oxidants, minerals, and vitamins. Banana pulp is soft, easily digestible flesh with simple sugars like fructose and sucrose that when eaten replenishes energy and revitalizes the body instantly. Because of this quality, bananas are being used by athletes to get instant energy and as supplement food in the treatment plan for underweight children. It also contains good amount of soluble dietary fiber which helps in easy bowel movement and reduction of constipation. It contains health promoting flavonoids, polyphenolic antioxidants such as *lutein, zea-xanthin, ß and α-carotenes* in small amounts. These compounds act as protective scavengers against oxygen-derived free radicals and reactive oxygen species (ROS) that play a major role in aging and various disease development processes.

Table 7: Nutritional Composition of bananas

Component	Quantity per 100 g	Component	Quantity Per 100 g
Energy	116 Kcal	Niacin	0.7 mg
Protein	1.2 g	Folate	20.0 mcg
Fat	0.3 g	Calcium	17.0 mg
Carbohydrates	27 g	Phosphorus	36.0 mg
Ash	0.9 g	Potassium	36.0 mg
Dietary fibre	3.5 g	Magnesium	27.0 mg
Carotene	78 ug	Iron	0.9 mg
Riboflavin	0.8 mg	Zinc	0.13 mg
Vitamin C	9.0 mg	Manganese	0.266 mg
Thiamine	0.05 mg	Selenium	1.856 mcg

http://nutritiondata.self.com/facts/fruits-and-fruit-juices/1846/2; Gopalan *et al.,* 1985

6.1 Phytochemicals

Polyphenols are secondary metabolites of extremely diverse chemical structures, also present in large amounts in bananas, with a majority of the antioxidant located in the fruit peel, and then the pulp (Manach *et al.*, 2004; Shahidi and Naczk, 2004; Faller and Fialho, 2010). The scavenging activities of polyphenols have been reported to be mainly due to their oxidation and reduction properties, thus enabling them to act as singlet oxygen collectors and hydrogen donors (Babbar *et al.*, 2011). Polyphenols play a major role in a plant's innate defense mechanism. Their contents in plants differ significantly as a result of variations in cultivar, growing conditions, maturity state, processing, and storage conditions (Jaffery *et al.*, 2003; Faller and Fialho 2010). The pulp of raw banana contains catechin, epicatechin, epigallocatechin, gallic acid, and prodelphinidin dimer (Arts *et al.*, 2000; Pascual- Teresa *et al.*, 2000; Del Verde-Mendez *et al.,* 2003; Harnly *et al.*, 2006), with most of these compounds belonging to the class flavanols, while its peel, according to Gonzalez-Montelongo *et al.* (2010), contains a large amount of dopamine and catecholamines. Polyphenols, over the past decade, have been receiving increasing and considerable interest from food scientists, nutritionists and consumers due to their high antioxidant role and health-beneficial properties. These secondary plant metabolites have been implicated in the prevention of cardiovascular diseases, cancers and type 2 diabetes (Arts and Hollman 2005; Scalbert *et al.,* 2005). The major compounds found in sap of different banana accessions, like Musa balbisiana, Musa laterita, *Musa ornata*, and *Musa acuminata*, and some cultivars were apigenin glycosides, myricetin glycoside, myricetin-3-O-rutinoside, naringenin glycosides, kaempferol-3-O-rutinoside, quercetin-3-O-rutinoside, dopamine, and N-acetylserotonin. These compounds were reported to have influenced various biological activities (Pathovon *et al.,* 2010).

Dopamine is a strong water-soluble antioxidant identified in the popular commercial banana *Musa cavendishii*. Banana contain dopamine at high levels in both the peel and pulp. Dopamine levels range from 80-560 mg per 100 g in peel and 2.5-10 mg in pulp, have been reported even in ready to eat ripened bananas. It had greater antioxidative potency than glutathione, food additives such as butylated hydroxyanisole and hydroxytoluene, flavone luteolin, flavonol quercetin, and catechin, and similar potency to the strongest antioxidants like gallocatechin, gallate and ascorbic acid. Banana is thus one of the antioxidative foods (Kanawaza and Hiroyuki, 2000).

In a study conducted on antioxidant activity of water extracts of banana peel it was comparable to those of synthetic antioxidants such as butylated hydroxyanisole andbutylated hydroxytoluene. Among all isolated components ß-sitosterol, malic acid, succinic acid, palmatic acid, 12-hydroxystrearic acid, glycoside, the d-malic and 12-hydroxystrearic acid were the most active against all the Gram-negative and positive bacterial species tested (Mokbel and Fumio, 2005).

6.2 Health Benefits and Therapeutic Properties of Banana

Banana has several health benefits and curative properties for many of the non-communicable and degenerative diseases. Some of the important health benefits and curative properties are described below.

Anaemia: Because of its high iron content (0.9 mg/100g), bananas can stimulate the production of hemoglobin and helps in curing anaemia.

Blood Pressure: Banana is the richest source of potassium (422 mg/100g) making it the perfect food for helping to beat blood pressure.

Brain Power: Research has shown that the potassium-packed fruit can assist learning by making pupils more alert.

Constipation: High in fibre, including bananas in the diet can help restore normal bowel action, helping to overcome the problem without resorting to laxatives.

Depression: It is reported that people suffering from depression, felt much better after eating a banana. This is because bananas contain tryptophan, a type of protein that the body converts into serotonin - known to relax, improve mood and generally make one feel happier.

Hangovers: One of the quickest ways of curing a hangover is to make a banana milkshake, sweetened with honey. The banana calms the stomach and, with the help of the honey, builds up depleted blood sugar levels, while the milk soothes and re-hydrates the system.

Heartburn: People suffering from heartburns feel better after eating bananas, due to its antacid effect.

Morning Sickness: Snacking on bananas between meals helps to keep blood sugar levels up and avoid morning sickness.

Mosquito bites: Rubbing the inside part of banana skin was reported to reduce swelling and irritation due to mosquito bite.

Nerves: Bananas are high in B vitamins that help calm the nervous system.

PMS: Eatting banana can reduce the premenstrual syndrome because of its vitamin B6 which regulates blood glucose levels, they affect the mood swings.

Seasonal Affective Disorder (SAD): Bananas can help SAD sufferers because they contain the natural mood enhancer, trypotophan.

Smoking: Bananas can also help people trying to give up smoking, as the high levels of Vitamin C, A1, B6, B12 they contain, as well as the potassium and magnesium found in them, help the body recover from the effects of nicotine withdrawal.

Stress: When stressed, the metabolic rate rises, thereby reducing the potassium levels in the body. These can be re-balanced with the help of a high-potassium banana snack. Potassium is a vital mineral, which helps normalise the heartbeat, sends oxygen to the brain and regulates body's water-balance.

Strokes: The New England Journal of Medicine reported that eating bananas as part of a regular diet can cut the risk of death by strokes by as much as 40%.

Temperature control: In many other cultures bananas is considered as a "cooling" fruit that can lower temperature of expectant mothers. In Thailand, for example, pregnant women eat bananas to ensure their baby is born with a cool temperament.

Ulcers: The banana is used as the dietary food against intestinal disorders because of its soft texture and smoothness. It is the only raw fruit that can be eaten without distress in over-chronic ulcer cases. It also neutralises over-acidity and reduces irritation by protecting coating/ lining of the stomach.

Burns: Tender banana leaf is aseptic, soft and non-sticking due to which dressing the burn wounds with banana leaf is being practiced in some parts of India for faster and less painful healing of burns.

6.3 Validation of Health and Therapeutic Claims

Fruits are rich sources of a variety of nutrients, including vitamins, trace minerals, dietary fiber and many other classes of biologically active compounds.

These phytochemicals can have complementary and overlapping mechanisms of action, including modulation of detoxification enzymes, stimulation of the immune system, reduction of platelet aggregation, modulation of cholesterol synthesis and hormone metabolism, reduction of blood pressure, and antioxidant, antibacterial, and antiviral effects. Although these effects have been examined primarily in animal and cell-culture models, experimental dietary studies in humans have also shown the capacity of vegetables and fruit and their constituents to modulate some of these potential disease-preventive mechanisms. The human studies have relied on intermediate end points related to disease risk (Lampe, 1999). A hemoagglutining activity has been reported in some species of banana, such as Cavendish bananas (Davila, 2002). This agglutining activity can be attributed to the presence of lectins, which are proteins of non-immune character that can be combined specifically and reversibly with sugars (Sharon and Lis, 1989). Moreover, a strong toxic activity of aqueous extracts from some species of banana and other fruits has been demonstrated against *Artemia salina* (Sharon and Lis, 1989). The ulcer protective and healing effects of unripened plantain banana have been studied in various models of experimental gastroduodenal ulceration and in patients with peptic ulcer, through reports of mucin secretion, mucosal cell shedding, cell proliferation, AOA, glycoproteins, and prostagladin synthesis (Goel and Sairam, 2002). At present, there is overwhelming evidence that free radicals cause oxidative damage to lipids, proteins, and nucleic acids. Therefore, reactive oxygen species (ROS) such as O^{-2}, OH^{-1}, or lipid peroxyl radical (LOO) might lead to many biological complications, including carcinogenesis, mutagenesis, aging, and atherosclerosis (Halliwell and Gutteridge, 1989). Because of the large variety of pathologies that have been related to ROS, an extensive search for new antioxidants that could lessen the effects of ROS on organisms has begun (Toyokuni, 1999; Villa-Caballero *et al*, 1999; Pellegrini *et al.*, 2003).

Chapter – 7

Post Harvest Losses and its Causes

Different levels of post harvest losses in fruits and vegetables are quoted in literature, but much of this is unreliable because these are based on the estimates to a great extent rather than actual experimental trials. These available figures are estimates of loss during movement of banana and plantains from rural areas or plantations to urban markets. The quality of transportation infrastructure including the roads has a great bearing on the extent of losses. In developing countries, where bananas are mostly grown, these arrangements are inadequate. In 1978 the National Research Council of USA reported that the post harvest losses of bananas and plantains as 20-80% (NCR, 1978). In 1981 FAO reported losses of banana ranging from 0-100% (FAO, 1981). Being a perishable commodity, the banana require special care and treatment after harvest like pre-cooling, cold storage, refrigerated truck, etc. As per one of the recent estimations, it is reported that in India 4.18% postharvest losses occur during farm operations in banana and 2.42% during storage (Nanda *et al*., 2012). Though pre-harvest care plays a major role in shelf life and quality of fruits and vegetables, lack of proper postharvest care further deteriorates its quality and shelf life.

The post harvest losses occur from the time of harvest till they reach the consumer. In addition to physiological causes several other factors are also responsible for post harvest losses in banana and plantains. These include losses caused by mechanical injuries, biological factors like diseases and pests, environmental factors such as temperature, relative humidity and gas composition inside the storage rooms. Improper harvesting and handling practices, inappropriate packing and transport methods can all lead to post harvest losses. Certain pre-harvest factors also influence the post harvest deterioration of perishables like banana. Post harvest losses could be quantitative or qualitative or both. Quantitative loss occurs due to reduction in weight as a result of moisture loss and loss of dry matter by respiration. Qualitative loss is deterioration of freshness leading to loss of consumer appeal and loss in nutritional components like vitamins, minerals, and other metabolites/ phytochemicals.

There are different pre and post harvest factors responsible for post harvest losses, which could be genetic / varietal, physiological, mechanical, abiotic like temperature, humidity, water and nutrition or biotic like pathological and entomological. Besides, there are certain pre-harvest care factors also which affect the post harvest shelf life and quality of banana. The post harvest causes for losses include improper harvesting, unscientific handling and transportation, lack of storage facilities, spoilages and lack of processing facilities.

7.1 Pre-Harvest Factors Responsible for Post Harvest Losses

7.1.1 Genetic/Varietal

The genetic composition of banana greatly affects its quality including its storage life. Though banana breeding has been existing for more than seventy years (Ortiz *et al.*, 1995) future developments will depend on cultivar selection as the seeds of banana are highly recalcitrant. Varieties differ in many characteristics, including visual appearance (example size), yield and quality (Hofman and Smith, 1993; Kader, 1992). Variety also has an effect on total soluble solids, firmness and juiciness (Shewfelt and Prusia, 1993). The genotypic characteristics of any cultivar will also vary in response to environmental effects. As discussed in the varietal characteristics, most of the varieties like Grand Naine, Robusta, Dwarf Cavendish, Morris, etc., belonging to Cavendish group (AAA), have longer green life but shorter yellow life. In general the post climacteric deterioration is much faster in varieties having A genome compared to those having B genome like Ney Poovan (AB), Nendran (AAB), and Karpuravalli (ABB). Quality and yield are important varietal characteristics. With increase in yield potential of a variety the quality like total soluble solids decreased. Similarly with increase in edible portion content the TSS decreased Ara *et al.* (2011). The varietal characteristics were also reported to affect the quality of processed banana products like papad (Venkata Subbaiah *et al.*, 2013).

7.1.2 Physiological and Mechanical

The bananas contain high moisture content of about 75-80% making it more susceptible to spoilage. Being a climacteric fruit, it is subjected to ripening after harvest due to commercial considerations. During the ripening it is susceptible to microbial attack, as the starch converts to simple sugars making it readily available to the pathogens for its growth and multiplications. An increase in rate of respiration is one of the most important physiological changes contributing to shorter shelf life of banana. Concurrently an increase in the level of ethylene production is also observed in bananas contributing to faster ripening. An increase in transpiration due to respiratory heat contributes to loss in weight. The soft texture of banana pre-disposes it to mechanical damages during growth and development on the plant as well as during its postharvest handling and marketing. The mechanical injury inflicts wound on the surface of the fruit and the wound induced ethylene production hastens the ripening and senescence of fruit. These mechanical injuries could be caused during growth and development due to rubbing or during the harvesting operation due to knives or improper harvesting tools. Further it could caused due to faulty handling practices. Winds damage the fruits by causing abrasions due to rubbing of fruits against the bunch support or leaf petioles. These mechanically damaged fruits are more prone to spoilage during post harvest period and have shorter post harvest life. Post harvest physiological changes like ripening, respiration and weight loss and its interplay with storage environment conditions determine the quality and shelf life of banana (Wills *et al.*, 1998).

7.1.3 Abiotic Factors

Abiotic factors like high temperature some times induce pre-mature ripening in banana. The temperature in which the crop is grown can affect its quality and post harvest life. For most fruits the higher the temperature during the growing period, earlier the time of harvest. Exposure of fruits to high temperature on the tree could influence the response of those fruits to high and low post harvest storage temperatures. High temperature coupled with lesser water availability leads to poor fruit filling and lesser yields. Water deficiency at critical stages like flower bud differentiation, flag leave emergence and fruit development also affect the yield, quality and shelf life. If bananas are allowed to mature fully before harvest and harvesting is done shortly after rainfall or irrigation, the fruit can easily split during handling operations. This allows microbial infection and post harvest rotting. Heavy winds are common during May –June and Nov-Dec in peninsular India resulting in fall of plants. Unseasonal cyclonic winds due to depression also damage lots of banana plantations.

Pre-harvest nutrition also influence the post harvest shelf life and quality. Generally high levels of nitrogen confers poorer keeping quality. The form of nitrogenous fertilizers used also affect the quality. Banana is a potassium loving plant and it plays a significant role in its yield, maturity and quality. Higher potash, particularly in the form of sulphate of potash has shown to increase the keeping quality and yield compared to nitrogenous fertilizer. The potassium increases the total soluble solids and dry matter content, while reducing the physiological disorders. Micronutrients has been found to play a significant role in influencing the yield and quality of banana. Yadlod and Kadam (2008) observed significant effect of plant growth regulators and micronutrients on days to harvest from flowering and days to ripening from harvesting .Two sprays of 1% micronutrient resulted in early maturity and early ripening, while 80 ppm GA_3 delayed the maturity. The micronutrient mixture called 'Banana Special' or 'Banana Shakthi' was found to increase the yield, reduce the incidence of pests and diseases and improve the quality of banana by farmers in southern states of India. Deficiency of micronutrients like zinc, molybdenum and boron results in physiological disorder. 'Latundan' is one of the varieties of banana that is highly susceptible to finger drop once the fruit is already ripe. Finger drop is associated with the weakening and softening of the pedicel during ripening causing individual fingers to dislodge easily from the crown. This limits its shelf life and results in lower price. Esguerra *et al.* (2009) observed that 4% calcium chloride delayed the onset of finger drop by 2 to 4 days after the attainment of full yellow color and extened the shelf life of fruits. Ripening of the fruit treated with calcium chloride proceeded normally without any effect on the physico-chemical and sensory attributes at the ripe stage.

7.1.4 Biotic Factors

The biotic stress factors like insect damage (scarring beetle, pseudostem borer), fungal pathogen attack (pre-harvest leaf spot, fruit rot, cigar end rot) also deteriorate the quality and shelf life of banana. Improper harvesting and handling practices cause physical damage to the fruits resulting in induction of ripening and spoilages. Same effect is caused by the mechanical damages to the fruits during handling and transportation. Besides the pre-harvest diseases like Fusarium wilt, sigatoka leaf spot, cigar end rot, BSV, BBMV and BBTV, post harvest diseases like crown rot, anthracnose and Aspergillus rot also affect the quality and shelf life of banana fruits. Among the insect pests pseudostem borer, corm weevil and banana scarring beetle cause severe pre and post harvest losses in banana. The effected bunches show improper filling, pre-mature ripening and spoilages due to secondary infection. Among the nematodes banana root lesion nematode and root knot nematode make the plants stunted and results in poor filling and fruit quality.

7.1.5 Maturity

The stage at which physiological development ceases and the fruit is ready for harvest is called as maturity. The time required for maturation differs based on the variety and climatic conditions. In banana, the level of maturity at harvest also affect the shelf life and quality. The fruits harvested at early mature stage have longer storage life and better quality. While full mature fruits have the best quality in terms of sweetness, sugar /acid ratio and aroma, their shelf life is shorter with short pre-climacteric phase resulting in faster senescence and spoilage (Desai and Deshpande,1978). To offset this problem, the stage of harvest of banana bunch is decided based on the destination of market. Bananas meant for distant markets are harvested when slightly immature, compared to those for hinterland markets. However, fruits harvested too early (less than 75% level maturity) fail to ripen properly and fetch no price in market. Many plants topple or break due to heavy winds during the bunch maturation period and such bunches are wasted.

7.2. Harvesting Method

Faulty method of harvest results in injury to the fruit rendering it vulnerable to post harvest diseases and rotting. There are primitive ways to highly mechanized system of harvesting practices, across the world. In India the method of harvesting are primitive due to small farm size. In some places the bunches are harvested using a harvesting knife tied to a pole which is raised upto the level of bunch peduncle and a sharp cut is given with a jerk. The falling bunch is held by one or two people standing below the plant. They are then lowered to and kept on ground till all the harvest of farm is completed.

However, the large plantations like in Costa Rica, Ecuador and Columbia are amenable for mechanized harvesting and bunch handling. Careful harvest, collection and field packaging, can reduce the losses and maintain the quality.

7.3 Post-Harvest Causes Responsible for Losses

7.3.1 Improper Handling and Transportation

The traditional method of post harvest handling of bananas in India is very primitive. No care is taken to prevent injury to the fruits during transportation from field into the truck. The fruits are loaded into the trucks directly without dehanding and cushioning. Though some green banana leaves are provided at base and sides of the trucks, it does not take care of transportation shock. As these bunches undergo multiple loading and unloading operations till it reaches the consumer, the fruits are subjected to bruises, rubbing and cracks. These damages become evident after ripening, in the form of blackened skin and softened portions of pulp. These soft spots undergo faster deterioration leading to spoilage and loss.

7.3.2 Lack of Storage Facilities

In the domestic trade the banana fruits are not stored at low temperature. Under ambient conditions the injury/wound induced ethylene liberation takes place from the fruit triggering its ripening. The damaged portion becomes vulnerable to the attack of post harvest spoilage organisms. Being a climacteric fruit, banana starts senescing fast once the ripening has started. The warm and humid tropical conditions aids the spoilage of such fruits.

7.3.3 Microbial Spoilages

The banana fruits being succulent are liable to damage and deterioration during harvesting, transportation, marketing, storage and consumption, if they are not handled properly. The deterioration mainly results due to physical injuries and enzymatic action by the attack of microorganisms. There are several reports of post-harvest diseases of banana from different countries (Cordeiro and Matos, 2005; Jones and Stover, 2000; Llyas *et al.*, 2007; Sulali *et al.*, 2004; Sadhu *et al.*, 1992). Anthracnose caused due to fungus *Colletotrichum musae*, is very common post harvest disease affecting banana, characterized by black spots on the surface of ripe fruit. Crown rot caused by *Botryodiplodia theobromae*, Fluffy white rot by *Fusarium moniliformeand* Cigar-end rot caused by *Verticillium theobroma* are the common post harvest diseases encountered in bananas. All these diseases renders the fruit unmarketable and leads to loss.

7.3.4 Lack of Processing Facilities

All the climacteric fruits have short shelf life and are susceptible to loss due to various reasons. Being perishable commodities, no method is available for its storage beyond 1-2 months. Under such circumstances, processing is essential to prevent the losses. The processed products have very long shelf life beyond two months or even upto one year. Lack of processing facilities at the catchment areas is one of the many limitations due to which the post harvest losses are huge in banana.

The inadequate support infrastructure which is the biggest bottleneck in expanding the food processing sector in general and banana in particular. The bottlenecks include fragmented supply chain, inadequate cold storage and warehousing facilities; poor road, rail and port infrastructure. Also, lack of modern logistics infrastructure such as logistics parks, integrated cold chain solutions, last mile connectivity, dependence on road over rail, customized transportation, technology adoption (barcoding, RFIDs) are some of the lacunae that exist in supply chain & logistics sector in India.

Chapter – 8

Pre-Harvest Factors Affecting the Shelf / Storage Life and Quality

Various pre-harvest factors like climatic conditions during fruit set and maturation, pest and disease attack, nutrition, soil moisture content affect the quality of fruits. The optimum temperature for higher yields and optimum quality is in the range of 25-35°C as there is an ideal balance between assimilation and leaf area increment (Chundawat *et al.*, 1983a & 1983b). Extremes of temperatures results in partially filled fruits either due to chilling or heat-stress injury. High relative humidity encourages virulent spread of leaf spot disease resulting in poor yield. As the pre-harvest and post harvest environment of banana and plantains affects their quality and marketable life, it is important to understand these factors and their interactions. In this section the effect of pre-harvest factors on post harvest quality and shelf life are discussed.

8.1 Climatic Conditions

The edible bananas are cultivated commercially in tropical to sub-tropical regions, between latitudes 30°N and 30°S. A suitable banana climate is a mean temperature of 80°F (26.67°C) and mean rainfall of 4 in (10 cm) per month. A dry spell of more than 3 months will adversely affect the banana crop.

8.1.1 Temperature

The atmospheric temperature during the growth and development influences the accumulation of dry matter content. It has been reported that the photosynthesis has a broad temperature optimum decreasing at an air temperature of les than 20°C and greater than 35°C (Turner and Lahav, 1983). Respiration increases exponentially with increase in temperature and the dry matter accumulation is a balance between the rate of carbon assimilation and the rate of respiration. That is the reason accumulated heat units are taken as one of the valid maturity index. As there is wide variation in this due to difference in the local conditions, it is location specific. Uma *et al.*, (1999) reported that bunches of banana cv.Jahaji showed the highest yield and good quality fruits when matured over the temperature range of 18-22°C (minimum), 28-26°C (maximum), relative humidity (68-83%) without rains which occurs during the months of September –October. Maximum yield and fruit quality in Malbhog and Ney Poovan were reported during January-April with a weather package of 18-26°C (minimum), 28-34°C (maximum) temperature, 71-76% RH and without rains. The quality parameters like total soluble solids and acidity are also influenced by season. Higher TSS and lower acidity was reported in Jahaji cultivar when fruits come to maturity during September-October The quality of fruits are the results of an interaction between the variety and prevailing weather conditions during fruit maturation. The highest yield in Malbhog variety was reported when the fruits matured in May-August, but the fruit quality in terms of higher TSS and lower acidity was recorded in summer maturing bunches (Uma *et al.,* 1999; Singh, 1976). Ney Poovan (a diploid variety) yields round the year under mild weather conditions with highest yields noticed during September to March months. Under extreme weather conditions, where the summers are very severe, the fruit yields were less, when bunch maturation occurred in May-August, but quality was better. Bananas harvested during the hot rainy season had the lowest pulp dry matter and total soluble solids. In optimal conditions without constraints other than temperature, a total cumu-lative temperature of 900 degree days is required to obtain the theo-retical optimum harvest stage of dessert bananas (Chang *et al.*, 2000). Bananas harvested during the cool dry season had the highest concentration of citric acid. Carbohydrate (dry matter, total soluble solids, and glucose and fructose) concentrations in fruit pulp decreased with an increase in mean daily temperature from bunch emergence to harvest (Christophe *et al.*, 2009). Also, a relationship between natural production conditions and physical fruit traits (texture and colour) were reported by Bugaud *et al.* (2007). The peel of green bananas harvested during the hot humid season was not as hard as that of bananas harvested during the cool dry season. In ripe bananas, a decreasing correlation was also noted between the mean daily temperature and the fruit yellowness. This interaction could be responsible for the yellower colour banana pulp and of bananas harvested during the coolest seasons. The green life of bananas harvested during the hot humid season was shorter than that of bananas harvested during the dry and intermediate seasons (Bugaud *et al.,* 2007).

The effect of high temperature is the function of time in which at 48°C the damage occurred in 3 h and 45 min, at 57°C it took 7 min and at 62°C it took only 1 min (Gowen, 1995). Under peel discolouration, the fruit was observed with air temperature of 33°C to 35°C but was associated with high light intensity and moderate to low rainfall. A more widely occurring effect of high temperature is bleaching of the chlorophyll in both leaf and fruits. Another important effect of temperature is chilling damage during crop development. It includes reddish brown streaking on the skin of the fruit, reduced latex flow, slow colour development of the fruit during ripening, a dull or grayish yellow colour in ripe fruits and brown skin in advanced stages. More subtle changes are reduced production of volatiles during ripening and disturbances to the evolution of CO_2 and ethylene (Gowen, 1995). In addition, Crane *et al.* (2005) reported that a temperature below 16°C but above 0°C resulted in failure of the flowering stalk or fruit bunch to emerge from the pseudostem (called choking) and an increase in fruit rotting during ripening. Young plants grown at sub-optimal temperatures produce thicker leaves, so they intercept less light and therefore have a lower relative growth rate (Van der Ploeg & Heuvelink, 2005).

In order to understand the influence of production systems and weather parameters on the development of Crown Rot of Banana (CRB) three farms with integrated production systems, and three organic farms located in the Caribbean region of Costa Rica, were selected and sampled every two weeks for a year. Results showed lower presence of CRB in the organic system than in the integrated system. The highest incidence occurred in periods of high precipitation and high relative humidity. Moreover, these weather variables correlated with an increase in CRB in the integrated production system farms. Increased frequency of rainfall, particularly exceeding 100 mm, correlated with increased CRB in the organic system farms. Regression analysis showed that incidence of CRB was associated with the type of system (integrated/organic), the cumulative average temperature, the highest accumulated temperature, the interaction between average cumulative temperature and days with rain, and with number of days with rainfall events. As the number of days with precipitation events increased and the average temperature increased, there was a decline of CRB. However, after 67 consecutive days with rainfall events, increased temperature produced increased CRB (Umana-Rojas *et al.*, 2011).

8.1.2 Light

Leaves absorb light to fix the atmospheric carbon dioxide and together with mineral nutrients from the soil, the dry matter is formed which is then partitioned among different plant parts. The photoperiod and quality of the light influence developmental processes and ultimately affect the yield and quality. An increase in light intensity was found to increase yield but not fruit quality in terms of fruit size and fruit soluble solid content in tomato (Verheul, 2012). Plants are known to benefit from equal distribution of light throughout the canopy; e.g., increasing the penetration of natural light into the canopy enhances productivity (Aikman,1989). Supplemental light has been shown to increase not only the amount of yield, but also the external and internal quality of many vegetable crops. For example, higher percentage of first class fruit and higher skin chlorophyll and dry matter content in cucumber (Hao and Papadopoulos, 1999), higher sugar content and ascorbic acid concentration in tomato (Dorais and Gosselin, 2002), and increased head firmness of lettuce (Gaudreau *et al.*, 1994) have been reported as results of supplemental light. In another study, according to the visual grading of fruit colour and structure during the storage, lighting regime had a small effect on the keeping quality of cucumber fruits. When the percentage of marketable fruits is considered, inter-lighting seemed to increase the keeping quality in the autumn–winter and spring crop (Hovi-Pekkanen and Tahvonen, 2008).

8.1.3 Relative Humidity

The atmospheric relative humidity directly affects the transpiration rate by altering the stomatal conductance. In winter the stomata becomes unresponsive to humidity. Besides this direct effect, the temperature and humidity determines the incidence of spread of leaf spot disease. The extent of leaf spot disease affects leaf lamina and the photosynthetic rate, thereby influencing the dry matter accumulation and quality.

Fruit texture, particularly firmness, is a critical quality attribute determining postharvest quality (Gross *et al.*, 2002). The firmness is dependent on the cell wall structure of the tissue (Brett and Waldron, 1996), and cell turgor which undergo an extensive change during ripening process. The most important postharvest process responsible for degradation of this cell wall structure is fruit ripening (Wakabayashi, 2000). The importance of water status for fruit firmness also reveals itself through the reversible physical effect temperature has on firmness as seen for apple and kiwifruit (Chen, 1993; Jeffery and Banks, 1994; Johnston *et al.*, 2001). With increasing and decreasing temperature, the water inside the fruit expands and contracts in volume. An increased cell tension due to increased turgor or temperature will increase the stiffness and the elastic modulus of the tissue (Chen, 1993). As viscosity of the cell walls will increase with decreasing temperature, the cell walls might become more brittle resulting in an increased stiffness but reduction in the cell wall strength. The overall effect will be the result of the net effect of temperature on both tissue tension and cell wall viscosity.

8.1.4 Wind

Wind is detrimental to banana plants. While light winds may shred the leaves thereby interfering with the photosynthesis; stronger winds may twist and distort the crown damaging the crop. Winds during the bunch bearing stage will topple the plant if it is not propped. Winds of 60 mph or over will uproot entire plantations, especially when the soil is saturated by rain. Windbreaks are often planted around banana fields to provide some protection from cold and wind. Cyclones and hurricanes are devastating, damaging the entire crop. Hurricanes has been one of the main reason for the shift of large scale banana production from the West Indies to Central America, Colombia and Ecuador.

The direct effect of excessive wind is the tearing of leaf laminae resulting in reduced yields. Indirectly it is involved in creating a greater vapour pressure deficit in the vicinity of leaf resulting in more transpiration. Excessive transpiration pulls more water from the soil and may set in motion a cascade of physiological processes. Normally excessive wind coupled with dry weather adversely affect the fruit yield. The fruits growing under such conditions are found to have more total soluble solids due to concentration of dry matter. In most of the bananas growing areas, the planting time is adjusted to escape the windy season during bunch bearing stage. However, some varieties of bananas and some times due to unpredicted weather conditions many plants in bearing lodge causing huge loss to the farmers, as the immature fruits fail to fetch any price in the market.

8.2 Water

Water is one of the most important limiting factors in cultivation of banana. Water is the medium for uptake of many minerals from soil, as well as a medium of translocation of photosynthates. Water management is essential to produce maximum amount of fruit or to produce fruit of a particular desired quality. Maximum yields were obtained with 290 to 300 kg N ha-1 and soil moisture tension close to field capacity during the entire crop cycle, but the maximum economic benefit (benefit/cost ratio) was obtained with 25 kPa- 200 N. Reduction (0.521 mg ha-1) in yield was linear for every unit (kPa) that moisture tension moved away from 10 kPa in case of banana cv.Grano Enano (Romero and Zamora, 2006;). Water deficit is one of the factors associated with the postharvest qualities of the fruit. The net effect of water deficit is reduction in the sugar/acid ratio (Turner, 1995). Inadequate irrigation to banana leads to delayed flowering, bunch size, delayed maturity of reduced fingers, induces physiological disorders and also poor keeping quality of fruits (Lopez, 1998). However, excess water also has detrimental quality consequences for plant. The photo-synthetic rate decreases with overly high water availability and low transpiration rates. High moisture content in fruit also tends to dilute the soluble solids leading to low flavor intensity. During dry periods, soluble solids accumulate rapidly inside plant structure. When the water supply increases suddenly, by irrigation or rainfall, water moves quickly to the plant cells in response to the osmotic gradient. The rapid rise in hydro-static pressure can rupture cell membranes and walls, leading to splitting of fruits and other structures (Shewfelt and Prussia, 1993). Furthermore, a high relative humidity during fruit development shortens the storage life and increases the incidence of finger drop and crown rotting (Munasque *et al.*, 1990). Generally, the water regime in the early stage of growth had a more pronounced effect both on dry matter accumulation and nutrient uptake than it did after bunch emergence (Gowen, 1995).

8.3 Spacing

Optimum plant density for banana is derived from complex integration of many factors like cultivar, soil fertility, sucker selection, management of crop, weeds, wind speed, topography and economic considerations (Simmonds, 1966). The plant density exhibited a significant effect on the quality of banana fruit. In an investigation, the plant raised under low density exhibited superior fruit quality in terms of T.S.S., reducing sugar, total sugar and sugar acid ratio (Chaudhuri and Baruah, 2010). They had a tendency to decrease with increasing plant density. However, a reverse trend was observed in case of titrable acidity and non reducing sugar. Higher acidity in higher plant population is probably due to shade effect where sugar conversion from organic acid is hampered due to lack of sufficient light. Elsiddig *et al.* (2009) observed that the wider plant spacing resulted in the highest exportable yield as compared to the closer spacing but the spacing had no significant effects on neither TSS nor taste of ripe fruits. Athani *et al.* (2009) observed maximum plant height (172.05 cm) and yield, (28.00 t/ha) in closely planted Rajapuri banana, while, the wider spacing recorded minimum plant height (163.70 cm) and yield (11.67 t/ha). Similar observations were also recoreded by Oluwafemi (2013) in Nigeria. Sarrwy *et al.* (2012) observed that banana plants grown at close spacing had taller pseudostem than plants grown under wide spacing. The highest yield per feddan (1.032 acres) was obtained from plants spaced at 3x2 m with two plants per hole, followed by those at 3x1 m, with one plant per hole since it was 34.07 and 30.33 tons in the first ratoon and 34.80 and 31.50 in the second ratoon for both planting distances, respectively. Earliest bunch shooting and minimum days for harvesting were recorded with planting distance 3x4 m with three plants per hole and bunch emerged earlier (12-13 days) than bunches produced from plants spaced at 3x1 m with one plants per hole in both first and second ratoons, respectively. Heaviest bunches were harvested from plants at 3x4 m spacing with three plants per hole. The highest finger weight, length, total soluble solids (TSS%) and total sugars% were recorded from plant spaced at 3x4 m with three plants per hole. Hazarika and Mohan (2007) did not find any significant effect of spacing on the bunch weight of Jahaji banana. The plant densities did not have any adverse effect of on TSS of fruits in Jahaji, Gaint Governor, Basrai and Lacatan (Reddy and Patil, 2007; Chattopadhyay *et al.*, 1980; Chundawat *et al.*, 1982). There was no adverse impact of increasing planting density on quality of banana (cv.Rasthali Pathakapoora) fruits like TSS, sugars and acidity. Higher doses of nitrogen and potassium increased the total soluble solids and sugars but decreased the acid content in the ripe fruits (Dinesh Kumar *et al.*, 2008). In banana cv. Robusta, Mandal and Sharma (1999) reported that plant height, days required from flowering to fruit maturity and fruit yield were significantly increased in crop density. However, high density of 1600 and 2000 plant/ha resulted in a decrease in the number of fruits, number of hands per bunch, fruit weight and bunch weight. Nalina *et al.* (2000) studied influence of reduced NPK fertilizer application under high density planting in banana cv. Robusta (AAA) involving three to four suckers per hill.

The results of the experiment conducted by Kumar *et al.* (2008) clearly indicated that plant height (315.99 & 302.15 cm) and fruit yield (37.707 & 35.28 t/ha) increased significantly due to increasing planting density in main and ratoon crop. But, earlier shooting (by 10 & 15 days) was observed due to lower planting density (2,500 pl/ha). There was no adverse impact of increasing planting density on quality of banana fruits like TSS, sugars and acidity.

8.4 Crop Nutrition

Many horticultural crops are heavy removers of nutrients and high yields can only be sustained through the application of optimum doses of nutrients in balanced proportion. Banana crop is a great lover of nutrients and it exhausts macro and micronutrients from the soil in large quantities. Split application of inorganic fertilizers along with organic manures increases the leaf nutrient status, uptake and distribution of nutrients and ultimately it results in better growth, yield and fruit quality. Under tropical conditions, soil nutrients will be leached rapidly due to various factors and therefore it should be applied in small doses at shorter intervals (Thangaselvabai *et al*, 2009). Mineral nutrients form an important part of plants and play a key role in plant growth and development. Impaired growth and development not only affect the yield but also the shelf life and quality. Croucher and Mitchell, (1940) observed that the application of N (150 g/ plant) along with P_2O_5 and K_2O increased plant growth and yield in banana cv. Gros Michel. According to Ramanathan *et al.* (1973) N @ 55 g /plant was sufficient to get profitable yield in cultivar Poovan. A fertilizer dose of 180 g N, 180 g P_2O_5 and 180g K_2O applied in three split doses within six months of planting enhanced the growth and yield of Basrai banana (Chundawat *et al.*,1983a). Mostafa (2005) observed the highest bunch weight, number of hands/bunch and number of fingers/bunch with best finger quality (weight, length and diameter) when the nitrogen and potassium was applied at different intervals throughout the year. Fruit yield, quality and biomass increases with high level of N and P (Lahav, 1972; Kohli *et al.*, 1976; Ram and Prasad, 1988). A fertilizer dose of 500g N, 250 g P_2O_5 and 200 g K_2O along with 0.02% S was found to be optimum for obtaining profitable yield in banana cv. Grand naine (Mostafa and Abd El- Kader, 2006). The increase in yield, dry matter content and chemical characteristics of fruits with nitrogen application may be due to enhanced photosynthesis which led to the accumulation of more carbohydrates and other metabolites ultimately translocating towards the fruits tissue (Hegde, 1988; Mustaffa 1988;). Higher nitrogen content in the leaves leads to more accumulation of phosphorus and it ultimately increased the meristematic growth, leaf area index and induced protein synthesis (Jauhuri *et al.*, 1974; Parida *et al.*, 1994). Significant improvement in TSS content and pulp/peel ratio of fruits was observed due to NPK nutrition in banana (Pandey *et al.*, 2005).

Among the major nutrients, potassium not only improves yields but also benefits various aspects of quality. It is well known fact that nitrogen plays an important role in development of foliage which in turn would affect the fruit yield. Potassium is often referred as the quality element for crop production (Usherwood, 1985; Ramesh Kumar *et al.*, 2006). The crucial importance of potassium in quality formation stems from its role in promoting synthesis of photosynthates and their transport to fruits, grains, tubers, and storage organs, and to enhance their conversion into starch, protein, vitamins, oil etc. (Mengel and Kirkby, 1987). Plants require potassium for the production of high-energy molecules (ATP) which are produced both in photosynthesis and transpiration processes. Potassium maintains the balance of electric charges in chloroplasts, which is required for ATP formation. Potash starvation either during early vegetative stage or during shooting and post shooting stage results in significant yield and quality reduction (Beena *et al.*, 1993; Mahalakshmi and

Sathyamoorthy, 1999). Potassium stimulates early shooting, increase in number of hands, finger size, improves quality and sweetness. It also increases keeping quality. High potassium and calcium will give high dry matter and glucose content in the peel and pulp of banana fruit (Caussiol, 2001). The sulphate of potash is more responsive than murate of potash in banana. Respiration was lower in potassium deficient plants and therefore the main effect of low potassium supply on dry matter production would be through reduction of photosynthesis. On the contrary, low levels of nitrogen, phosphorus and magnesium give high dry matter in the pulp. Low potassium produces thin fruits and fragile bunches. Excess potassium over nitrogen causes a premature ripening of fruits (yellow pulp). In addition, a potassium supply above that which influence growth and yield, changes in reducing and total sugars have reported. As potassium supply increases, the amount of sugar will increase and the acidity decreases. Thus potassium supply has an effect on fruit quality over and above its influence on yield (Gowen, 1995). The fruit growth is restricted by two ways due to low potassium levels. The translocation of carbohydrates from leaves to fruit is reduced and, even when sugars reach the fruit, their conversion to starch is restricted (Martin-Prevel, 1973). It has been observed that as the potassium level increases, the sugar/acid ratio increases because of increased sugars as well as reduced acidity (Vadivel and Shanmugavelu, 1978). Beneficial effects of K include high juice content, high vitamin C content, uniformity and acceleration of ripening of fruits and resistance to bruising or physical breakdown during shipping and storage (Ganeshmurthy *et al.*, 2011). Potassium also is reported to improve fruit weight and number of fruits per bunch and increases the content of total soluble solids, sugars and starch (Bhargava *et al.*, 1993). In an experiment conducted to study the response of banana Cv. Mysore (AAB group) to the addition of potassium, it was revealed that potassium affected fruit characteristics (fruit length, diameter and weight) as well as the fruit yield of Mysore banana (Weerasinghe and Premalal, 2002). Sugar content increased from 11 to -4 13.1 per cent by soil application of 480 g K per plant to Dwarf Cavendish banana. Mustaffa (1988) found increase in acid content of Hill banana with increasing dose of K from 0 to 250 g K per plant and ascorbic acid from 62.2 to 108.6 mg/100 g pulp by 300g of K_2O. Low potassium nutrition results in thin and fragile bunches with shorter shelf life (Von Uexkll, 1985).

It is considered that nitrogen requirement is second only to potassium in terms of amount needed for growth and production (Twyford, 1967). As the banana plant cannot store nitrogen, it is always in short supply to the plant (Butler,1960). In the absence or shortage of soil nitrogen, it is re-distributed from old banana leaves to the young ones (Vorm and Diest, 1982). Because of this the leaves exhibit a deficiency symptoms characterized by pale green colour of leaves and midribs and petioles becoming reddish-pink (Murray 1959). The resultant effects are decreased dry matter accumulation and maturation period. Nitrogen affects bunch weight through its effect on number of hands and fingers, fruit circumference or caliper grade and average fruit weight (Arunachalam *et al*, 1976). It also affects the fruit quality by reducing total sugars, total soluble solids and increased acidity (Mustaffa, 1983). But these results are controversial to the observations of Hazarika and Mohan (1990) and Adonis *et al* (2009). Adonis *et al.*, (2009) found that level of N and K did not affect the diameter of fruits, acidity and total soluble solids. After the first cycle, regardless of N rates, the application of K_2O rates reduced the pulp firmness. In the ratoon crop, K_2O rates showed significant interaction with N, and the largest yield was obtained with application of 1600 kg ha^{-1} of K_2O.

Micronutrients especially iron, manganese, zinc, boron, copper and molybdenum are very essential for various activities like chlorophyll synthesis, enzyme action, uptake of nutrients, photosynthesis, oxidation reduction reaction, respiration and nitrogen metabolism etc. The effect of micronutrients in enhancing the various metabolic processes, growth and yield parameters of banana was studied by several workers (Srivastava, 1964; Twyford and Walmsely, 1968; Walmsely and Twyford, 1975; Mahouachi, 2007).

Calcium and magnesium are the two secondary elements required in banana nutrition. Application of dolomite ($MgCO_3$) and magnesium sulphate ($MgSO_4$) improves growth and yield of banana. Calcium nutrition plays a significant role in the uptake of nitrogen and micronutrients (Walmsely and Twyford, 1975; Vorm and Diest, 1982; Reghupathi *et al.*, 2002). Calcium deficiency leads to inferior fruit quality and the skin split of ripe fruits are common (Charpentier and Martin-Prevel, 1965).The beneficial effect of magnesium and sulphur in improving the leaf chlorophyll content, growth, yield and quality of banana was reported by HosamEl-Deen, (2002) and Mostafa and Abd El- Kader (2006).

Twyford and Walmsely (1968) recommended an annual application of 30 Kg $MgSO_4$, 28 Kg iron citrate, 12 Kg borax, 3Kg $ZnSO_4$ and 1Kg copper sulphate/ ha for correcting the nutrient deficiencies and improving the yield . Foliar application of $ZnSO_4$ 0.5%, $FeSO_4$ 0.2%, $CuSO_4$ 0.2% and H_3BO_3 0.1% during 3, 5 and 7 months after planting increased bunch weight and TSS in ripe fruits.

8.5 Pests and Diseases

Pests and diseases has caused a decline in production of banana in several parts of the world. The banana weevil *Cosmopolites sordidus* Germar (Coleoptera: Curculionidae) is a major insect in all the banana and plantain growing areas of the world (Waterhouse and Norris, 1987). Detrimental effects of weevil infestations include: premature plant death, stunted growth, delayed fruit maturation, production of small bunches, reduced number of suckers, reduced sucker vigour and development of water suckers (Abera *et al.*, 1999).

Among the nematodes attacking the banana the most damaging species include: *Radopholus similis* (Cobb) Thorne (Kashaija *et al.*, 1994), *Pratylechus coffeae, P. goodeyi* (Namaganda *et al.*, 2000) and *Helicotylechus multicinctus*. Nematodes attack roots, causing lesions thereby reducing water and nutrient uptake to the upper parts of the plant and also paving way to pathogenic micro organisms.

Banana Bacterial Wilt is caused by a bacterium called *Xanthomonas campestris* pv. *musacearum*. The disease affects all types of bananas and the major characteristics of the disease are yellowing and complete wilting of the plant starting with the most peripheral leaves (Tushemereirwe *et al.,* 2003). The most common symptoms are wilting and premature ripening of the bunch in addition to male bud wilting and sometimes discoloration. The fruits of affected plants have poor quality and shorter storage life.

Fusarium wilt (Panama disease), caused by a soil borne fungus *Fusarium oxysporum* Schlecht. f.sp *cubense* (E.F. Smith) Snyd. and Hans. is a typical vascular disease causing disruption of translocation and systemic foliage symptoms in bananas, which eventually lead to collapse of the crown and pseudostem (Jeger, *et al.*, 1995).

Black Sigatoka leaf spot caused by *Mycosphaerella fijiensis* Morelet and yellow sigatoka caused by *Mycosphaerella musicola Leach* can cause yield losses upto 50%. Poor fruit filling and premature ripening of fruits affecting the quality and shelf life are the common post harvest problems encountered with this disease.

Banana Streak Virus (BSV) causes necrotic leaf streaks, lethal stem necrosis, pseudostem splitting and choking in which fruit bunches may fail to emerge from pseudotems. Severely damaged plants may have reduced bunch size and misshapen fruits. The affected fruits are not suitable for marketing.

Banana crown rot disease caused by *Lasiodiplodia theobromae* and *Colletotrichum musae* is a common post harvest disease in banana. Besides fungicidal dips natural compounds like plant oils (*Ocimum sanctum,*

Cymbopogan citratus, Cynmbopogan nardus and *Cymbopogan martinii* oils) completely arrested the mycelial growth and quality loss and extended the shelf life in bananas (cv.Robusta-AAA) during storage (Sangeetha *et al.*, 2010; Abeywickrama *et al.*, 2004). Other environmental friendly integrated treatment strategies include treatment with alum+basil oil (0.16% or 0.20% v/v) which reduced the crown rot disease without affecting the physico-chemical qualities like titratable acidity, total soluble solids, pH, and fruit firmness (Abeywickrama *et al.*, 2009).

For the management of banana anthracnose, antifungal effects of Arabic gum (AG) (5, 10, 15 and 20%), chitosan (CH) (1.0%), and the combination of AG with CH were investigated in vitro as well as in vivo. CH at 1.0% and 1.5% had fungicidal effects on *C.musae.* AG alone did not show any fungicidal effects while the combination of 1.0% CH with all tested AG concentrations had fungicidal effects. In vivo analysis also revealed that 10% AG incorporated with 1.0% CH was the optimal concentration in controlling decay (80%), showing a synergistic effect in the reduction of *C. musae* in artificially inoculated bananas (Maqbool *et al.*, 2010; Ali *et al.*, 2012).

Hot water treatment (HWT) delayed the ripening and prolonged the shelf life in Bungulan bananas. HWT lessened crown rot severity by 50% after 7 days and 33% after 14 days (Alvindia, 2012).

Banana fruits treated with biocides formulated from essential oils of aniseed, coriander or black cumin seeds were found to reduce the decay and extend the storage life of banana. Further, biocide treated banana fruits kept their good quality for longer storage periods compared to non-treated fruits, due to lower ripening rates which promisingly prolonged shelf-life. Biocide application retarded the conversion of starch into simple sugars, especially at the low temperature rate. Additionally, the imposed treatments maintained vitamin C in banana pulp and lowered the decline in peelchlorophyll content (Mahmoud *et al.*, 2010).

Fuzzy pedicel is a new postharvest disease of banana, affecting single fingers. It is caused by *Sporothrix sp.* and several species of *Fusarium* (Tarnowski *et al.*, 2010).This disease was found to affect when fruits are stored at 25°C.

8.6 Bunch Covering

Proper bunch management technologies, particularly bagging the bunches with polyethylene covers have been found to improve the yield and quality of banana (Robinson and Nel,1984). Plastic bunch cover should be placed around the bunch of fruit when the last hand is visible. The sleeves come in different sizes, thicknesses and colours. Covers protect the bunch from being damaged by birds, scarring beetles and helps to increase the weight and improve the quality of the fruit. For centuries, in New Guinea, the ancient practice was to use old banana leaves for wrapping around maturing bunches. In 1936 it was demonstrated that covering bunches with hessian protected them against winter chilling and improved fruit quality. Later, paper bags were used to a limited extent. In 1949 and the early 1950s, plastic covers came into existence. They were immediately successful and now covers are extensively used in banana cultivation all over the world. Besides protection from pests and diseases there are several advantages of bunch covers like it increases the yield, brings about uniformity in size or grade, protects from damages of insects, bats and birds and improves the colour and look and the overall quality.

Covering of bunch has helped in Cavendish and Nendran banana to get attractive colour and better quality. In order to reduce the build up of temperature inside the sleeves, appropriate ventilation is provided. The level of ventilation in bunch covers have been reported to affect the physical fruit qualities like length, girth and weight of fingers. The highest harvest index was observed in bunches having bunch covers with 6% ventilation (Narayana, *et al.*, 2007; Singh and Bhattacharya, 1991; Debnath *et al.*, 2001). Covers made of PVC was advantageous in areas where temperature was a limiting factor (Ganry, 1975). The colour of the sleeves have not been uniformly found to affect the quality of fruits. Stevenson (1976) found no effect of colour of sleeves on fruit filling in summer, but in winter it speeded up filling of bunches. Covering banana bunches with blue polyethylene sleeves a fornight prior to harvest ensured uniform colour development and avoided the rupture of skin. White bunch covers were found to significantly increase the total chlorophyll content in fruit skins (Choudhury *et al.,* 1996). Though thickness of the covers did not significantly affect the quality of fruits, based on its re-usability, 0.5mm thick perforated polybags are normally recommended. The bagging is done when all hands have emerged. The fruits in covered bunches were found to be more attractive, free from blemishes, early to mature with increased size of fingers (Parmar and Chundawat, 1984; Robinson and Nel, 1984; John and Scott, 1989).

8.7 Maturity and Maturity Indices

Maturity is a stage at which the physiological development of a fruit ceases, and the natural process of ripening is initiated. The bananas are harvested at a specific maturity stage in which the maturity indices are based on the age of the bunch, the interval between flowerings and harvesting, the filling of the fingers or the colour of the skin and pulp and brittleness of the flower end of the bunch. The filling of the fingers is the criterion mostly used. This standard is typically completed by other visual criteria like the evolution of the peel colour of the fruits. Most of these criteria depend on the cultivar. If the filling of the fingers was combined with the colour of the pulp, evaluating harvest maturity could become more objective (Chang *et al.*, 2000). Determination of stage of harvest is of paramount importance in climacteric fruits like banana, as the shelf life of fruits depends on the stage of maturity at harvest. In banana, the initiation of ripening is marked by a rapid rise in respiration rate of the fruit reaching a peak after which it falls as ripening progresses. During long distance transportation of bananas, it needs to be kept green and unripe to prevent the loss of quality. Hence they are harvested green, transported and the ripening is initiated artificially at the destination market. Too mature fruits ripens immediately and may split during handling, particularly if irrigated before harvest. Depending on the distance of the market, the stage of harvest is determined. For marketing locally, it is harvested at full maturity, and for distant markets it has to be harvested at 75-80% maturity. Though there are several harvest indices for determining the stage of maturity, none of these are universally practiced as it varies depending on the variety and various other pre-harvest practices. Heat units, age of the bunch after emergence from pseudostem, angularity of fingers, peel to pulp ratio, length of the finger, diameter of the finger or caliper grade, firmness of the fruit, brittleness of the floral end are some of the indices worked out by researchers. According to the study of Ahmad *et al..* (2007), speed of ripening was independent of harvest maturity at the ripening temperature of 16°C but that it was reduced at 13°C for more immature fruit. Also chilling injury occurred at 13°C which reduced the acceptability of the fruit. The symptoms of chilling injury progressed as the fruit ripened and were more apparent at colour stage 6 than early colour stages. Bananas from the early stage of maturity showed increased levels of chilling injury symptoms. Early harvesting resulted poor quality ripe fruit in terms of the total soluble solids and eating quality. Therefore, the variation in the quality of ripe fruit on a commercial scale could be due to the difference of harvest maturity.

In India, where more than 99% of bananas produced are consumed domestically, the fruits are harvested at full maturity. However, in exporting countries like Ecuador and Colombia, the bunches are marked immediately after emergence from the pseudostem using colored ribbon. The color of these ribbons are changed weekly so that the age of bunch can be easily known by color of the ribbons. In these countries, Grand Nain or Valery cultivars of banana become ready to harvest for export in 98-112 days (Gowen, 1995), while in India, it attains this stage by 90-100 days. Besides, this the experience of the harvester is also used to discriminate and assess the maturity. The effect of the maturity on quality and shelf life has been studied by various workers.

8.7.1 Heat Units

As temperature in the banana growing area has a significant role in fruit maturation, maturity can be assessed based on heat units accumulated. Ganry and Meyer (1979) developed a formula based on the relationship between grade (G) and temperature (T), G=15.7 + 0.202 ST, where 'ST' is the sum of the average daily temperature above 14°C. This formula was developed with historical weather data to predict long-term and short-term harvest dates, with ± 7days accuracy. Based on this the cultivar Willams required 900 degree days to attain a caliper grade of 34 mm from the time of last hand emergence. As this is based on the mean temperature during the bunch development period, it varies from one season to another even for the same variety and same locaton. Stover and Simmonds (1987) reported that shooting to harvest varies from about 80 days in hottest months to 120 days in the coolest month for AAA clones and upto 180 days for ABB clones in tropical climates.

8.7.2 Shooting to Harvest Period or Age after Emergence of Flower

When a bunch is harvested based on caliper grade alone, a mixture of fruit with varying maturity as wide as 50 days can be seen in the same box. Such consignments are called 'mixed ripe' or 'ripe in transit'. It is due to this reason, the bunches are tagged using various coloured ribbons.

8.7.3 Angularity of Fingers

As the fruit matures the visible angles of the individual fingers disappear and become less prominent. Fruits are classified as three quarter, full three quarter and full mature based on angularity. However, disappearance of angularity cannot be solely relied on, because the cooking cultivars remain angular even at full maturity.

8.7.4 Caliper-grade

Though harvest maturity depends on the specification provided by the importing country, calibration size or caliper grade and length of finger are commonly used indices. For determining the caliper grade, the middle finger in the outer whorl of the second hand is callipered at the thickest part of the finger. The grade is expressed in three ways depending on the country as- total thirty seconds of an inch. Fruits with calibration of 40/32 to 48/32 are considered mature for harvest (Anon, 1988).

8.8 Maturity vs Shelf life and Quality

Maturity plays an important role in deciding the time of harvest as it effects the quality and shelf life. Higher correlation coefficients were obtained with age than with caliper grade. The green life of hands was seven days for over mature fruits and 56 days for immature hands when stored at 13°C in Ecuador. Desai and Deshpande (1978) observed that fruits harvested at early mature stages had a longer storage life and better quality. Krishnamurthy (1989) reported a longer shelf life of Robusta banana when harvested early (at 100 days after flowering). The quality of the banana improves with increase in maturity. After ripening, compared to the early harvested fruits, the later harvested ones were reported to have higher percentage of total soluble solids, total sugars, reducing sugars and less starch and acidity (Dhua *et al.*, 1992).

8.9 Pre-Harvest Sprays

Banana receives various kinds of sprays during its cultivation which includes sprays of fungicides/insecticides, growth hormones, foliar liquid fertilizers and other chemicals. Spraying of tonics or growth promoters widely available in market under different brand names are sprayed by farmers to improve the yield and quality of fruits. All the kinds of sprays have long lasting effects even after harvest. The incidence and intensity of pre-harvest diseases like leaf spot or sigatoka can affect the finger filling initially and later the fruit quality after ripening. Most of the post harvest pathogens of banana find their dormant or nascent existence on the fruits during pre-harvest period and spreads during storage and ripening. The diminishing self defense of the fruit during ripening favours faster spread of the disease. The effect of some of these sprays on shelf life and quality of banana fruit is briefly described below.

The gibberlins are known to promote cell elongation, thereby contributing to increase in the size of fruits like grapes and banana. Spray of KGA3 on the bunches two weeks after flowering was found to increase the yield (Lockard, 1975) and retard the ripening. Pre-harvest spray of gibberlic acid retarded the ripening and increased the total sugars. The response to pre-harvest spray of gibberlic acid is dependent on maturity of the fruits, as younger fruits showed better response than older bunches of banana (Mishra *et al.*, 1981). Application of GA3+7 about a month prior to harvest was found to be four times more effective in delaying the ripening by 16 days in Dwarf Cavendish. It is also reported to increase the pulp to peel ratio in banana cv.Gaint Governor. It is believed that this effect is brought out by increasing the levels of endogenous auxins and cytokinin and delaying the ripening processes (Gottreich and Halevy, 1982; Chellappan, 1983; Pradhan *et al.*, 1988). Parmar and Chundawat (1984) found that 100 ppm NAA, 20 ppm Kinetin with 1% mineral oil or Navras (a foliar formulation by private company) when applied at full inflorescence emergence delayed the fruit ripening in Basrai banana. In Nendran banana, out of various pre-harvest sprays like gibberlic acid, Biozyme, Navras, potassium nitrate and potassium sulphate at different concentrations, it was reported that potassium sulphate (2%) as two sprays was only found to be influence bunch weight, number of hands, number of fingers, finger length and girth. The quality parameters like TSS, acidity, total sugars and starch were not affected by any of these sprays (Mustaffa *et al.,* 2007).

Calcium is an important apoplastic (in plant cell walls and membrane) and symplastic (as a second messenger) nutrient (Zocchi and Mignani, 1995). Calcium affect the marketability of the product, as it affects the quality parameters like attractiveness, organoleptic, nutritional value, intact state and shelf life. (Jha, 2006). Philippe *et al.* (2010) did not find any effect of calcium fertilization or spray on the shelf life or quality of Cavendish banana.

When 'Ardhapuri' banana was sprayed twice with micronutrients 1%, the fruits matured early (106 days), while IAA @80 ppm delayed the maturity. Maximum weight of bunch (22.76 kg), number of hands (9.25) and fingers (133.25) were recorded in IAA 80 ppm spray. Maximum length (21.95 cm) and girth (16.20 cm) of finger was recorded in micronutrient mixture 1% with two sprays (Yadlod and Kadam, 2008).

Finger drop is a serious problem in post climacteric stage of banana during ripening. Essguerra *et al.* (2009) reported this as a serious problem in Latundan variety of banana. Among various treatments, 4% calcium chloride was found to delay the onset of finger drop by 2 to 4 days after the attainment of full yellow color. This treatment resulted also in the lowest incidence of finger drop. However, it did not have any effect on ripening, physico-chemical changes or sensory attributes of banana cv.Latundan. Gibberellic acid and ethanol though slightly retarded the ripening, they did not control finger drop.

Anthracnose, crown rot and blossom end rot are common and serious postharvest diseases of banana. Anthracnose, caused by the fungus *Colletotrichum musae*, is a latent infection. The fungal spores infect immature banana in the field but symptoms appear only at the ripening stage. Naturally occurring antagonists on host surfaces are a promising component of biological crop protection. Being a latent infection, anthracnose is a postharvest disease which is difficult to control by treatments at the postharvest stage (Janisiewicz and Korsten, 2002). Integrated methods of biological control with fungicides as field sprays have shown effective control of anthracnose of mango and avocado (Korsten *et al.*, 1997).

8.10 Harvesting

The harvesting operation is one of the stages at which the fruits suffer injury and damage. A right method of harvest, depending on the produce, is essentially required to prevent mechanical damage and injury to the fruit. In banana inclusion of dirt to the fruit occurs at this stage, as most of the farmers after harvest keep the bunches on ground in the orchard or field, before being moved to the collection centre. In tall varieties like Saba, Karpuravalli, Monthan and even Poovan, improper method of harvest can damage the fruits due to impact injury when the bunches fall on the ground. Some times the harvesting tool like sickle or knife can accidentally inflict wound on fruits. These injuries become site of invasion for pathogens.

Chapter – 9

Post Harvest Factors Affecting The Shelf Life and Quality

The act of detaching the fruit from its parent plant is called harvesting. As seen in the preceding chapter the stage and method of harvest affects the quality and shelf life of bananas. After harvest the protection provided by the plant to the fruit ceases and the harvesting act itself gives a kind of shock accompanied by a physical injury to the fruit. This triggers a series of biochemical changes in the fruit ultimately leading to its senescence and death. The post harvest shelf life of the fruit is dependent on the surrounding environment immediately after harvest. The important post harvest factors affecting the post harvest life are temperature and relative humidity of the atmosphere, physical damage / bruises, microbial load on the surface of the fruit, packaging, transportation and ripening method.

9.1 Temperature

It is a well known fact that the rate of deterioration of perishables increases two to three-fold with every 10 °C increase in temperature. Temperature has a significant effect on how other internal and external factors influence the commodity, and dramatically affects spore germination and the growth of pathogens (Kader and Rolle, 2004). It is for this reason that a produce intended for long distance market should be pre-cooled immediately and refrigerated at its optimum storage temperature, so that all the biochemical changes occurring in response to the harvesting act are reduced at once and the subsequent rates of these changes are slowed down. However, exposing the fruits to too low temperature below its optimal can result in freezing or chilling injury, while exposure to high temperatures can cause heat injury. Delays between harvesting and cooling or processing can result in direct losses due to water loss and decay and indirect losses such as those in flavor and nutritional quality. Kader *et al.* (1978) found that delays of 24 h at 30 and 40°C between harvest and processing of tomatoes result in in 5 and 12% loss in ascorbic acid respectively. Similarly the quality in terms of ascorbic acid reduced in chill injured bananas (Miller and Heilman, 1952; Lee and Kader, 2000).

9.2 Relative Humidity

Relative humidity (RH) is defined as the moisture content (as water vapor) of the atmosphere, expressed as a percentage of the amount of moisture that can be retained by the atmosphere (moisture holding capacity) at a given temperature and pressure without condensation. RH can influence water loss, decay, incidence of some physiological disorders, and uniformity of fruit ripening. Condensation of moisture on the commodity (sweating) over long periods of time is probably more important in enhancing decay than is the RH of ambient air. An appropriate RH range for storage of fruits is 85 to 95 percent while that for most vegetables varies between 90 and 98 percent.

9.3 Physical Damage

Bruising significantly affected the physical quality, chemical composition and storage life of fruits. Various type of physical damage (surface injury, impact, bruising) are major contributors to deterioration. Mechanical injuries are not only spoil the appearance but also accelerate water loss, stimulate higher respiration and ethylene production rates, and favour decay. Decay is one of the most common cause for post harvest deterioration as many microorganisms usually follows mechanical injury or physiological breakdown of the commodity which allow entry to microbes.

9.4 Microbial Load

Most of the post harvest pathogens find their way from field. As the fruit is nakedly exposed in the atmosphere, they gather dust, secretions / excretions of visiting or attacking insects or/and other pre-harvest sprays. If these microbes are not washed away after harvest, they gain strength and attack the fruits which lost its active defense mechanism after harvest and are vulnerable to pathogens. Some times fruits exhibit pre-harvest cracks due to various reasons and these cracks harbour pathogenic microbes.

9.5 Packaging

Use of improper packaging material or lack of packaging of fruits during transportation can cause injury during loading and unloading. These injuries, though superficial can impair the physical appearance of fruits as they turn black after ripening. Unclean packaging material (particularly re-used several times) can harbor the pathogens and other chemical contaminants which gain access to the fruit during handling and render the fruit unsafe for consumption.

9.6 Transportation

Mostly the roads in the rural areas are not well laid or maintained due to which the fruits suffer damage during transporation. Often the trucks loaded with banana bunches are covered with tarpaulin sheets to protect from rains during transportation. These trucks are not fitted with refrigeration system. Hence, the field heat as well as the respiratory heat along with humidity builds up in the truck resulting in initiation of ripening of fruits. Due to lack of ventilation the humidity also increases and favors microbial growth. The vibratory shock of transportation while traveling on damaged roads injure the fruits resulting in atleast 10% drops or finger shattering before reaching the destination. The shatter fingers are not marketable and contributes to post harvest losses.

Lack of facilities for long distant transportation of bananas discourage farmers and traders in inter-state trade. This compels the farmers to sell the fruits locally even at throw away price. Though banana special rail wagons have been introduced for transport of bananas from Maharashtra to Delhi, the undue delay in order to get the prescribed load causes loss as these wagons are not refrigerated and it takes longer duration to reach Delhi than the trucks. The green wagons in superfast trains are of limited capacity which facilitate transport of only seasonal assorted fruits and vegetables in small quantities.

Chapter – 10

Post Harvest Physiology, Shelf Life and Quality Changes

Banana like many other fruits and vegetables contains high moisture content, due to which they have relatively high metabolic activity and makes it easily perishable. High perishability of banana pose a major challenge in its marketing as fresh fruit. Any strategy to manage the perishability or extend the shelf life demands a thorough understanding of the physiological changes it undergoes after harvest. Any fruit or vegetable has three distinct phases in its life viz., growth, maturation and senescence. Maturation usually may commence before growth ceases, and after attainment of maturity there is hardly any increase in the dry matter content accumulation. Senescence is defined as a period when anabolic (synthetic) biochemical processes give way to catabolic (degradative) processes, leading to aging and finally death of the tissue. Ripening is akin to fruits, which indicates the commencement of senescence. One of the major changes exhibited by climacteric fruits like banana is ripening and its associated changes in colour, texture, aroma and taste. These changes are preceded by a rapid rise in rate of respiration.

10.1 Respiration

Respiration is a catabolic process wherein the stored organic materials like carbohydrates, proteins, and fats, are broken down into simple end products with a release of energy. In the normal respiratory process oxygen is consumed and carbon dioxide is released and during this process. During respiration there is a loss of stored food reserves which leads to hastening of senescence, as the stored food is exhausted. Also, there is loss of flavor quality, especially sweetness. The rate of deterioration of harvested commodities is generally, proportional to the respiration rate (Irtwange, 2006, Hailu *et al.*, 2013).

Bananas are harvested when they are still green (pre-climacteric stage) and are ripened artificially using a ripening aid. Fruits that are allowed to ripe on the tree often split, and tend to be mealy. During the pre-climacteric phase, mature green fruit have a low basal respiration rate and ethylene production is almost undetectable. This period is also called the "green life". It is desirable to have a long green life for the fruits intended to be exported, as it ensures lesser damage and loss of quality. Green life can be extended by decreasing temperature to 14°C, and storage under low O_2 ($\leq$8%) and high CO_2 ($\geq$2%) (Marriott and Lancaster, 1983). Since banana is a climacteric fruit it exhibits a respiratory peak during ripening. Within a couple of days, respiration rate of the hard green banana fruit may rise to about five times at the climacteric peak and then fall as the ripening advances. During this period there is considerable loss of water due to transpiration (Salunkhe and Kadam, 1995). The respiratory heat also enhances the water loss due to transpiration. This cumulative loss of reserved food (due to respiration) and the respiratory water loss leads to weight losses in the product resulting in a decreased value (Kays, 1997).

As the respiration progresses the oxygen level in the storage environment is depleted and if it is excessively depleted, anaerobic conditions occur that rapidly spoil the fruits. Simultaneously, the carbon dioxide released during respiration builds up in the storage environment. Excessive build up of carbon dioxide leads to carbon dioxide injury and some varieties of bananas are sensitive to carbon dioxide. Therefore, periodic ventilation is essential to replenish the depleted oxygen and remove excessive carbon dioxide. The type and design of packaging material determines the duration of ventilation required in the storage chamber (Kays, 1997; Hailu *et al.*, 2013). The two major respiratory substrates found in fruit are sugars and organic acids. Both sugars and organic acids are found largely sequestered within the vacuole, and form a major contribution to the overall flavor of the fruit and the balance between sugars and organic acid can markedly affect the taste of the fruit. Both sugars and organic acids originate from photosynthetic assimilates. Fruits differ in how they assimilate and accumulates during development and ripening. Some fruits have accumulated the bulk of their carbohydrate prior to the onset of ripening. This is either stored primarily as starch, as in banana, or as sugars, as in tomato. Fruits which accumulate their assimilate prior to ripening can be harvested at the mature green stage and still attain acceptable flavor on ripening.

This is important when considering the need for early harvest to optimize shelf life (Tucker, 1993).

10.2 Effect of Ethylene on Ripening

The rate of ethylene production in climacteric fruits like banana increase with maturity at harvest, physical injuries, disease incidence, increased temperatures up to 30°C and water stress (Pesis, 2004). On the other hand, reduction in the level of ethylene production by fresh fruits are caused by low storage temperature, reduced O_2 levels, and elevated CO_2 levels around the commodity (Irtwange, 2006). Exposure of climacteric fruits to ethylene gas advances the onset of rise in respiration rate and climacteric peak.

Though it is well known that ethylene is intimately involved in the initiation of ripening in banana, its mode of action is unknown (Seymour, 1993). Unripe banana show a constant, but low level of ethylene production until the onset of ripening. Ethylene production then increases and is followed by a rise in the rate of respiration. During the ripening process, ethylene production reaches the peak while the rate of respiration is still increasing. As the rate of ethylene production declines, the rate of respiration reaches its maximum at around 125 mg CO_2 Kg-1 and then declines slightly, but remains at high level (Seymour, 1993). Therefore, in the trade of banana, to achieve optimum fruit quality, postharvest techniques like controlled atmospheric storage and modified atmospheric packaging are used in order to modulate the physiological processes of ripening.

10.3 Transpiration Loss

According to the study of Irtwange (2006) transpiration is the evaporation of water from plant tissues. Water loss is a very important cause of produce deterioration, with severe consequences. Water loss is, first, results in loss of marketable weight and then adversely affects appearance (wilting and shriveling). Also, the textural quality is reduced by enhanced softening, loss of crispness and juiciness, and reduction in nutritional quality. The nature of the epidermal layer of the commodity governs the level of water loss which is also affected by environmental factors. Eventually, transpiration is a result of mor-phological and anatomical characteristics, surface-to-volume ratio, surface injuries and maturity stage on the one hand, and relative humidity, air movement and atmospheric pressure on the other hand. As a physical process, it can be controlled by applying waxes and plastic films as barriers between the produce and the environment, as well as by manipulating relative humidity, temperature and air circulation.

The transpiratory loss from banana skin continues even after the bunches have been cut. The magnitude of transpiration depends on temperature and relative humidity and in trend it shows a similarity to ripening. The green fruit, immediately after cutting, shows an initial fall in transpiration rate and then settles down to a steady level depending upon temperature and humidity; at the climacteric there is a sharp peak followed by the attainment of a new steady state (Simmonds, 1959).

There is usually a final rise in water loss which is related to degenerative changes of the skin caused by fungal attack; since the skin in this stage is senescent the loss can hardly be described as transpiration. These facts were established for individual fruits; because the bunches ripens progressively from the top hand to the bottom, the climacteric transpiration peaks of individual fingers are concealed and whole bunches (or bigger bulks of fruit) show simply a steady rising curve of water loss during ripening (Simmonds, 1959).

10.4 Postharvest Biochemistry of Banana

The ripening process, at the fruit level involves several biochemical pathways like degradation of starch to sugar, change in the peel and pulp colour, cell wall changes, change in the concentration of volatiles and acids (Gowen, 1995).

10.4.1 Texture

Most fruit soften during ripening and this is a major quality attribute that often dictates shelf life. Fruit softening could arise from one of the three mechanisms: Loss of turgor; degradation of starch; or breakdown of the fruit cell walls. Loss of turgor is largely a non-physiological process associated with the postharvest dehydration of the fruit, and as such can assume commercial importance during storage. Degradation of starch results in a pronounced textural change, especially in those like banana, where starch accounts for a high percentage of the fresh weight (Tucker, 1993; Turner, 2001).

10.4.2 Carbohydrates

Starch forms about 20 to 25% of the fresh weight of the pulp of unripe bananas. During ripening this starch is de- graded rapidly and the sugars sucrose, glucose and fructose accumulate; traces of maltose may also be present. Sugars are present in the green fruit only in very small amounts, average about 1 to 2% of the fresh pulp; they increase to 15 to 20% at ripeness, the beginning of the increase coinciding with the respiration climacteric. Starch disappears concurrently, dropping from about 20 to 25% in the green fruit to about 1 to 2% in the ripe fruit; it is higher in the ripe plantain (6%) than in the dessert bananas (Simmonds, 1959). In the banana pulp sucrose is the predominant sugars, at least at the start of ripening, and its formation precedes the accumulation of glucose and fructose. The peel tissue also contains starch, about 3% fresh weight, and appears to show similar changes in carbohydrate during ripening. These characteristic patterns of carbohydrate metabolism can be altered under certain environmental conditions such as exposure to elevated temperatures during ripening (Seymour, 1993; Turner, 2001).

10.4.3 Pigments

There is a decrease in chlorophyll content from between 50 to 100 ìg per gram fresh weight to almost zero in ripe fruit, while caroteniod levels (xanthophylls and carotene) remained approximately constant at 8 μg per gram fresh weight. There is reduction in total caroteniod content in the peel during the early stage of ripening followed by caroteniod biosynthesis at the yellow-green to yellow-ripe stage (Seymour *et al.*, 1987).

10.4.4 Cell wall change

Softening of fruits appears to be closely linked with changes in their cell walls structures. In bananas, the changes in texture of the fruit during ripening probably result from alterations in both cell wall structure and the degradation of starch (Seymour, 1993).

Hemicellulose and pectin are components of cell wall and their concentration during ripening follows a trend similar to that of starch, dropping from 7 to 8% of the fresh pulp in the green fruit to about 1% at ripeness. There are indications of interconversion of starch and hemicellulose in the early phases of storage of green fruit. In behavior, however there is an important difference from starch, for the hemicellulose disappear whether or not the fruit is normally ripened; in fruit kept in prolonged cool storage, starch concentration declines slowly so that a high starch content is characteristic of "chilled" fruit until a very advanced stage of ripeness. The hemicelluloses disappear at the normal rate under such treatment and their fate is still mysterious. They may be connected with loss of dry weight of the fruit which exceed the loss that can be accounted for by respiration (Simmonds, 1959).

10.4.5 Organic acids

Pulp pH and total titratable acidity are important postharvest quality attributes in the assessment of fruit ripening quality. Generally, when fruits are harvested at matured green stage, the pulp pH is high but as ripening progresses pH drops. Thus, the pulp pH could be used as an index of ripening (Dadzie and

Orchard, 1997). The skin of the fruit shows similar trend but is slightly delayed with respect to the pulp; this is not surprising in view of the fact that ripening proceeds from the core of the pulp outwards. Various measurements of pH range between about 5 to 5.8 for the pulp of the green fruit and between about 4.2 and 4.8 for post climacteric fruit (Simmonds, 1959).

The levels of organic acids in a fruit can markedly affect its taste. Generally, in banana the pulp shows an increase in acidity during ripening and the main organic acid present are malic, citric and oxalic. As ripening advances, acidity declines presumably due to the utilization as respiratory substrates. The astringent taste of unripe bananas is probably attributed at least partly to their oxalic acid content, which undergoes significant decarboxylation during ripening probably by the action of oxalate oxidase (Seymour, 1993). Despite the increase in acid with ripening, is less that the rate of increase of sugars and total soluble solids (Gowen, 1995; Abbas *et al.*, 2012).

10.4.6 Enzyme Activity

Banana fruit contains several hydrolytic and oxidative enzymes. The relative activities of α-mylase, starch phosphorylase, acid phosphate, and catalase increased considerably in banana fruits stored for 5 weeks at 20°C. The stage of maturity of banana fruit at harvest significantly influenced most of the physical and biochemical constituents and the activities of some enzymes. Banana fruits harvested at mature or early mature stage had a longer storage period and better quality. The potential storage life of banana fruits decreased with an advance in stage of maturity (Salunkhe and Kadam, 1995).

10.5 Postharvest Pathology of Banana

Postharvest diseases can cause serious losses of fruits both in terms of quantity and quality. Fruits infected with disease have no market value. There are many postharvest diseases of banana, cooking banana and plantain. Among these important ones are crown rot, anthracnose and cigar-end rot (Dadzie and Orchard, 1997).

10.5.1 Crown Rot

Crown rot is one of the most important postharvest diseases of banana/plantain. It is characteristically a disease complex caused by several fungi, sometimes in association with other micro-organisms such as bacteria. The most common pathogens associated with crown rot are *Colletotrichium musae* (*Gloesporium musarum), Fusarium roseum, Fusarium semitectium and Botryodiplodia theobromae.* When the hands are cut from the stems, the massive open wound is an ideal weak spot for crown rot fungi to enter and grow. Fungal spores on the fruit in the field are carried along (after harvesting bunch) to the packing house. Spores follow the fruit right into delatexing baths, where they are drawn deeply into the weak spot, the wound on the crown tissue (due to dehanding). Spores also remain on the outside of the fruit and are packed (Dadzie and Orchard, 1997; Dionisio, 2012). Symptoms of crown rot include softening and blackening of tissues at the cut crown surface, white, grey or pink mould may form on the surface of the cut crown and infected tissue turns black and the rot may advance into the finger stalk. The control of crown rot starts in the field with the regular removal of leaf trash. Proper field sanitation can greatly reduce the number of crown rot fungi spores present and not keeping rotting fruits or plant waste materials near the packing station. Dehanding should be done carefully with a sharp knife so as to avoid leaving a ragged cut. Finally, postharvest treatment of fruits with an effective fungicide is essential (Gowen, 1995; Dionisio, 2012).

10.5.2 Anthracnose

Anthracnose is one of the important postharvest diseases of banana, cooking banana and plantain. It is caused by the fungus, *C. musae*. Anthracnose is common on wounds, but it is capable of attacking sound fruits as well. Occasionally, it invades the necks of the fingers when they are damaged. It is characterized by numerous small circular and brown to dark brown spots (Gowen, 1995; Dadzie and Orchard, 1997). Preventive measures begin in the plantation. It is important to maintain strict hygiene or sanitation in the plantation and pack house, in order to minimize the number of spores available for infection. All cultural practices that reduce scarring and injury to the fruit will prevent anthracnose (Dadzie and Orchard, 1997).

10.5.3 Cigar-end Rot

Cigar-end rot of banana and plantain is an important postharvest disease caused by the fungi, *Trachysphaera fructigena*. Cigar-end rot is essentially a plantain disease (but it is also found in banana and cooking banana), the fruits being apparently most subject to attack in their more immature stages. The number of fingers affected in the bunch varies. The infection which starts with localized darkening and wrinkling of the skin originates in the perianths and spreads slowly backwards along the finger. The darkened area is bordered by a black band and a narrow chlorotic region which separates infected and healthy tissues. The principal method of control is frequent manual removal and burning of dead flower parts and infected fruits. Use of fungicide to control the disease is also recommended. In the pack house, care should be taken to cull infected fruits to avoid contaminating the washing water with spores. Cigar-end rot is effectively controlled by covering the flower (immediately after emergence) with a polyethylene bag before the hands emerge (Dadzie and Orchard, 1997). Some other lesser known post harvest diseases have also been reported in the banana (Sarkar *et al.*, 2013).

10.5.4. *Alternaria* rot

The infection starts as brown spots from the stalk end and advances towards the stylar end (Fig: 1).The spots turn to brownish- black with age. Repeated isolations from such spots revealed *Alternaria tenuissima* (Kunze) Wiltshire. in culture.

Corynespora rot: The infection starts as water soaked patches and spreads with time, usually develop at the tip of the fruit. These patches progresses towards the base resulting in softening of the fruit. Isolations made from such portions revealed *Corynespora cassiicola* (Berk. and Curt.) Wei. as the pathogen (Fig. 2).

10.5.5 Drechslera Soft Rot

The infection manifests as small circular, black to brown coloured spots near the stalk-end of the fruit. In advanced stages of infection bad odour along with plenty of mycelial mass covers the infected surface of the fruit. Repeated isolations from such infected portions revealed *Drechslera halodes*(Drechsler.) in culture (Fig. 3).

10.5.6 Brown-Blotch

The infection starts initially as dark-brown spots, which later coalesce to form bigger spots. Repeated isolations yielded *Fusarium chlamydosporum* (Wollenweber and Reinking) in culture (Fig. 4).

10.5.7. Fusarial Rot

The infected fruits show brown, sunken spots that slowly spreads along the surface and the fruit pulp turns soft and watery with the advancement of the infection. The tissue bit isolations and microscopic examination revealed *Fusarium pallidoroseum* (Cooke) Saccardo in culture (Fig.5).

10.5.8. Brown rot

The infection manifests as brown coloured spots on the fruit surface. These spots coalesce and spread along the surface of the fruit with age. Isolations made from such diseased tissues yielded *Fusarium poae* (Peck) WollenW. in culture (Fig. 6).

10.5.9. Macrophoma Rot

Infection starts as small, round, brown scattered spots on the fruit surface. These spots usually turn to dark-brown in colour and results in tissue disintegration. Isolation from such spots revealed *Macrophoma musae* (Cooke.) Berl.et Vogl. in culture (Fig: 7).

10.5.10. Charcoal Rot

Infection appears in the form of patches near the tip of the finger. The surface becomes slightly softened and blackened, extending towards the other part of the fruit. On severe infection dark colouration on the surface of the fruit appears with black to brown margin. Isolations made from such diseased portions revealed *Macrophomina phaseolina* (Tassi) Goid. in culture (Fig: 8).

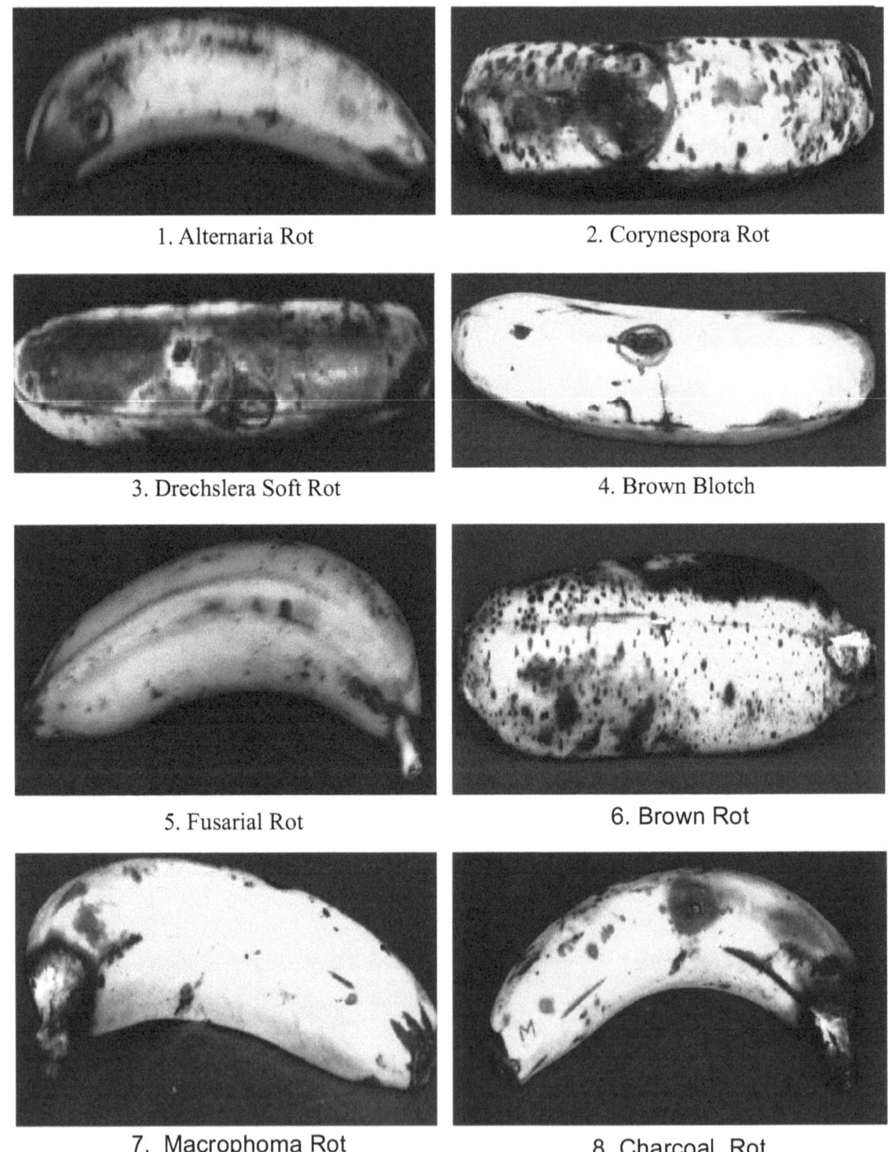

1. Alternaria Rot

2. Corynespora Rot

3. Drechslera Soft Rot

4. Brown Blotch

5. Fusarial Rot

6. Brown Rot

7. Macrophoma Rot

8. Charcoal Rot

Fig. 1 to 8

Chapter – 11

Improved Harvest, Postharvest Handling, Packaging and Transportation

The quality of fruits are influenced by many factors, of which the stage of maturity for harvest which is of prime importance depends on whether the fruit is meant for long distance transport or local consumption. Bananas to be marketed locally have to be harvested at a more mature stage than those to be transported over longer distances. Generally fully mature bunches are harvested for domestic market where fruits have to be ripened and consumed within a week. For regional markets where it may take about 2 weeks, the fruits are to be harvested at 90% maturity where in the angles are fairly well defined. For long distance transportation and export the harvesting is done at 75% maturity (Gowen, 1995) where the fruits have plainly visible angles.

11.1 Maturity Indices

Several parameters have been suggested for determining the maturity of banana fruits on the basis of their external appearance and chemical constituents at the time of harvest. But there are no universally accepted objective methods based on which maturity can be judged. Though lot of research has been carried out on this aspect, the results have been mostly location specific. These indices also differ considerably among different varieties and even strains of fruits (Gowen, 1995; Ogazi, 1996). Some of the indices used in different parts of the world are as follows.

11.1.1 Age After Shoot Emergence

The number of days from shooting to maturity is employed based on physiological age. Morton (1987) reported that 75% maturity is reached between 75-80 days after opening of first hand. It is necessary to have the data on days from shooting or bagging to harvest throughout the year. For this purpose in Ecuador and several other Central American countries the bunches are marked at the time of shooting by tagging with coloured stripes or ribbons and subsequently harvested at appropriate time taking into consideration other parameters like finger size and angularity. Bunches are harvested between 98 and 112 days in Grand Naine or Valery cultivars in southern Ecuador (Gowen, 1995). In tropics the average flowering to harvest duration varies seasonally from 98-115 days (Stover, 1979). Some studies indicated that plantain reaches proper stage for harvesting in about 90 days after shooting irrespective of time of planting (Sanchez Nieva *et al.*, 1971). The best practical and objective method identifying the harvest maturity is a combination of flowering to harvest duration, coloured ribbons and caliper measurement of finger diameter (Robinson, 1996). It has also been observed that best fruit quality is recorded when fruits are harvested between 12th and 15th week in Pisang Embun banana (Abdullah *et al.,* 1985). Wilson Wijerathnam *et al.* (1994) recommended that harvesting should be carried out at 11 weeks after flowering for cold storage and 12-13 weeks for air freighting of Embul bananas. At Burgershall, the time between flowering and harvesting varied from 110-204 days in 'Williams' banana (Robinson and Human, 1988). Banana cultivar 'Prata' harvested 105 days after anthesis ripen normally in storage. It was recommended that fruits should not be harvested later than 135 days after anthesis (Ayub *et al.*, 1996).

11.1.2 Angularity of the Fingers

Angularity is one of the visual methods of assessment of maturity. Visual inspection of the fullness of fingers or disappearance of angularity is one of the most commonly used methods (Gowen, 1995). Fruits are classified as three quarter (with clearly visible angles), full three quarter (less prominent angles) and full mature where angles virtually disappear. Angularity as a basis of maturity cannot be fully relied as cooking bananas have visible angles even at full maturity (Anon, 1988).

11.1.3 Diameter and Length of Fingers (Caliper Grade)

In Latin America, maturity is assessed using a specially designed spring caliper to measure the thickness of the middle finger of the outer whorl of the second hand from the distal end of the bunch. For export market, in most countries the caliper grade and length of fingers are the commonly used indices (Gowen, 1995; Anon, 1988).

11.1.4 Heat Units

In French Antilles, Ganry and Meyer (1979) found the following relationship between grade (G) and Temperature (T). G= 15.7 + 0.202 ST, where 'ST' is the sum of average daily temperature above 14°C. It required about 900 degree days for Cv. Williams to arrive at a grade of 34 mm from the time of last hand emergence.

11.1.5 Others

Several other signs are used as maturity indices like easy separation of flower ends, specific gravity, pulp to peel ratio, total soluble solids, solids to acid ratio, starch content, and even skin colour and fruit firmness. At various maturity levels of banana the fruit firmness values were calculated and it was observed that the peak value of 1.48 N/mm^2 was during the period from 120-125 days after fruit setting, which indicated appropriate period for harvest (Kharche *et al.,* 1996). Firmness can also be measured using Magness and Taylor pressure tester or Effigy electrometer (Ginsburg, 1970). But this is a destructive method and found no wide adoption. Non-destructive sonic techniques have also been developed for testing the firmness (Finney *et al.*, 1967). In India a minimum pulp to peel ratio of 1:1 has been accepted as being the acceptable level of maturity for ripening of Dwarf Cavendish bananas (Dalal *et al.,* 1970). However, by and large farmers use their ingenuity for deciding the maturity besides the market demand and prevailing price.

11.2 Harvesting and Field Handling

Banana bunches are harvested with a curved knife leaving about 6-9 inches of stalk attached to the bunch which serves as a handle for carrying. Improved handling methods have greatly reduced the bunch injuries. The methods of handling and harvesting practiced in Windward Islands during early 1970's used to result in heavy damages to fruit. Their technique involved dehanding the bunches either while still on plant or when they are cut and hung from a frame beside the plant (Robinson, 1996). The improved method of harvest followed is similar in pattern in most of the commercial plantations in the world. In this method a 'carrier' stands in position below the bunch to catch it and the 'cutter' cuts a notch into the pseudostem so that it falls slowly towards the carrier who guide the bunch on to a shoulder pad. The cutter severs the peduncle about 30 cm above the proximal hand and carries the bunch to a trailer or hangs on to a cable wire depending on the transport method used (Gowen, 1995; Narayana *et al.*, 2002a). In order to reduce the losses due to bruises and mishandling, cushioned pads of either PUF or blistered polyethylene sheets can be used. In several parts of India the method of harvest of banana involves giving a slant cut to the pseudostem and bending the upper portion of plant bearing bunch towards ground and severing the bunch from pseudostem leaving about 12-13 inches of peduncle above the first hand. The size of peduncle retained on the bunch varies depending on the size of the bunch. The bunches are then carried to a collection centre in field either on shoulder of a carrier person or in a wide basket on head (Narayana *et al.*, 2002a). In some cases the harvested bunches are carefully placed on 10-12 cm thick bed of banana leaves on ground, to allow the latex to flow from the cut ends. After latex flow ceased they are carried to the packing shed one or two at a time on a prototype of medical stretchers 150 cm long and 75 cm wide, made of thick smooth cloth. For weighing the bunches, a heavy string is tied to the stalk near the distal end and is weighed on a spring balance in field (Aradhya and Eipeson, 1999).

11.3 Packhouse Operations

11.3.1 Dehanding

Dehanding is done in a packhouse either in the field or at a different location. The packhouse essentially contains elevated cables onto which the banana bunches are hung without touching the ground. It will also have sufficient number of tanks for washing the banana hands, fungicidal treatment, drying tables to remove surface moisture, and packaging material. Dehanding of bananas is cutting of the hands from bunch stem. In order to avoid crown rot and finger drop it is essential to cut the hands with crescent like pad of tissue, which attaches the hand to main stem. Until 1960's Gros Michel, the then export cultivar of banana was exported as whole bunches. But later, when Cavendish group replaced it, the practice of dehanding came into vogue. Care has to be taken not to inflict slightest knife wound on the fruits or stalk of banana. The hands may further be cut into clusters having required number of fingers based on the size and shape of the packaging cartons. The latex from the cut end should not be allowed to drop or stain the fruits. The delatexing can be done physically by allowing the latex to ooze out or it can be floated in a tank containing detergent or sodium hydrochlorate solution 10 ppm where the latex gets washed out. While in the washing tank, the surface of the hands are scrubbed with a sponge to remove the dirt, floral remnants and other adhereing material. Later the clean hands / clusters are put into a tank containing fungicidal solution. The fungicide used should be the one which is approved or allowed. In India Benlate or Benomyl 500 ppm is allowed while for export Thaibendazole 100 ppm has to be used. The hands / clusters have to be held in the fungicidal solution for 3-5 minutes. Before using the chemical one has to ensure the acceptance of the fungicide in the importing country. In India for domestic market, dehanding is not commonly practiced except in Grand Naine. However, for export trade dehanding is essential during which time defective, damaged and under developed hands are culled out. After washing and fungicidal treatment the hands are dried under the blast of air from pedestal fans or blowers. Care should be taken not to leave any droplets of water on the surface or in between fruits (Narayana and Mustaffa., 2006).

11.3.2 Sorting and Grading

In recent past the Cavendish varieties like Grand Naine are dehanded to facilitate ripening by exposure to ethylene gas in ripening chambers. However, non-Cavendish bananas are still transported as bunches. For export trade dehanding is essential during which, defective, damaged and under developed hands are sorted and culled out. Selected bigger hands are graded based on the specifications (Codex Standards of grades) given below.

For export purpose 75-80% mature, green bananas are only suitable. Further after harvest the sorting and grading is essentially done to adhere to the minimum quality standards. The sorting should be done to remove immature or over mature and undersized hands, fruits with excessive scars, discolouration and any other form of mechanical damage. Deflowering of fingers, if any present, should be done before washing of the fruits. Additionally the specifications prescribed by the importing country with regard to the pesticide residues (MRL) have to be adhered to.

As per the Codex standards export quality bananas should be a minimum 14 cm long with 2.7 cm diameter. The reference fruit for measurement of length and girth is the medium finger on the outer row of the hand. The minimum length and girth of finger depends upon the customer requirement. Generally customer in international market demand for 7 inches length and 3.8 to 4.8 cm diameter. The general post harvest protocol followed for export of bananas is indicated in the following flow sheet of operations.

11.3.3 Codex Quality Standards for Export

A premium quality banana has to be very clean, free from decay, has an adequate finger length and diameter, does not have excess curvature, and upon ripening has the desired uniform bright yellow colour and sensory attributes of flavour and aroma.

- Bananas must be whole, firm, sound, clean and practically free from any visible foreign matter
- Practically free from bruises
- Practically free from pests and damage caused by pests
- With the stalk intact, without bending, fungal damage or desiccation
- With pistils removed
- Free from malformation of abnormal curvature of the finger
- Free from damage caused by low temperature
- Free from any strange smell

Grades specifications and tolerance limits within each grade are given in Table 8.

11.4 Packaging

In past, bananas for export from Fiji to New Zealand were detached individually from the hands and packed tightly in 33 kg wooden boxes. But this resulted in bruising of upper layer and the fruits in contact with sides. Reduction of fruit quality was found to offset the economic advantages of filling all the shipping space with fruits. Later wooden boxes were abandoned and suppliers resorted to packing fruits in cartons. Therefore, in the modern systems of handling and transport, quality control and good container design, have made carton packing not only feasible but also necessary. First the hands are graded for size and quality and then packed in layers in special ventilated cartons with plastic (PUF) padding to minimize bruising. In Indian banana marketing system the concept of packaging is yet to be practically adopted. A major quantum of bananas is traded either naked or just wrapped with banana leaves. No systematic grading and packaging is followed in bananas while it plays a very important role in marketing of the produce. As the consumer preference and buying behaviour is towards a standardized and fresh looking produce, appropriate packaging can lead to high price realization. Some research work carried out at CFTRI, Mysore and by the author at NRCB, Trichy has resulted in development of suitable CFB boxes for packing of banana both for export and domestic trade. The carton shall be a two-piece telescopic type made of virgin, 5 ply corrugated fibreboard with the dimensions of 65 X 30 X 25 cm. Each carton is provided with ~6% ventilation holes equally distributed on all four sides and side grooves for easy lifting of the carton. The banana hands are packed into the cartons in a lengthwise fashion with their cushion resting on the bottom of the carton and fingertips of the fruits pointing towards the upper lid. The cartons are packed with 6 to 7 hands (Cavendish variety), without any

Table 8: Grade Specifications

Grade Designation	Grade Requirements	Grade tolerance
Extra Class	➢ Bananas must be of superior quality. ➢ Characteristics of the variety and/or commercial type. ➢ The fingers must be free of defects, with the exception of very slight superficial defects, provided these do not affect the general appearance of the produce, quality, the keeping quality and presentation in the package.	Five percent by number or weight of bananas not satisfying the requirements of the grade, but meeting those of for Class I grade or, exceptionally, coming within the tolerances for that class.
Class I	➢ Bananas shall be of good quality. ➢ Characteristics of the variety and/or commercial type. The following slight defects of the fingers, however, may be allowed, provided these do not affect the general appearance of the produce, quality, the keeping quality and presentation in the package. - slight defects in shape and colour; - slight defects due to rubbing and other superficial defects not exceeding 2 sq.cm of the total surface area.	Ten percent by number or weight of bananas not satisfying the requirements of the grade class, but meeting those of Class II or, exceptionally, coming within the tolerances of that class.
Class II	➢ This includes bananas which do not qualify for inclusion in the higher classes, but satisfy the minimum requirements. ➢ The following defects may be there, provided the bananas retain their essential characteristics as regards the quality, the keeping quality and presentation. - defects in shape and colour provided the product remains the normal characteristics of bananas; - skin defects due to scrapping, scabs, rubbing, blemishes or other causes not exceeding 4 sq.cm. of the total surface area; The defects must not affect the flesh of the fruit.	Ten percent by number or weight of bananas satisfying neither the requirements of the class nor the minimum requirements, with the exception of the produce affected by rotting, major imperfections or any other deterioration rendering it unfit for consumption.

bruise. Between each layer of fruits a polyurethane foam sheet is provided as cushion, which avoids rubbing of hands against each other. CFTRI, Mysore and NRCB, Trichy have standardized bag in box or modified atmosphere storage with and without ethylene scrubber as suitable packages for export. This further enhances the storage life and also keeps the fruits in fresh condition (Aradhya and Eipeson, 1999; Narayana *et al.,* 2002a).

For different type of products, pack size, strength requirements and quantities dispatched by different types of transport need to be carefully assessed before deciding on the type of packaging. The need to reach distant markets particularly for exports and use of refrigerated transport has posed new challenges for developing appropriate packaging materials.

For domestic market, particularly short distance, after the packhouse operations the hands have to be placed in the ventilated plastic crates and the crates can be stacked one over the other in the truck. This will prevent damage to the fruits during transportation as well as prevent the piling up of waste (leaves, peduncles, *etc.*) in the market place. This will also facilitate easy ripening using ethylene gas.

11.5 Precooling

Field heat generated due to the harvest stress can cause overheating of the fruits. This may result in damage of the plant tissues and acceleration of the biochemical activities, causing spoilage. Rapid cooling after harvest is generally referred to as pre-cooling and particularly benefits rapidly respiring fruits like banana. Pre-cooling is essential as refrigerated ships, land vehicles and containers are not designed to handle the full load of field heat but are designed to merely maintain pre-cooled produce at the selected carriage temperature. Several methods like hydro cooling, air-blast cooling, vacuum cooling and liquid carbon dioxide cooling have been in use, separately or in combination, to take care of this heat load from horticultural crops. Hydro cooling is achieved by dipping/ drenching, rinsing/ immersing or spraying cold water over the banana bunches for effective field heat removal. In air blast cooling method, banana bunches are subjected to a jet of cooled air for removal of field heat whereas in vacuum cooling the fruits are treated under vacuum for few minutes.

Out of various methods of pre-cooling like air-cooling, hydro-cooling, contact ice cooling, vacuum cooling and evaporative cooling; hydro-cooling and air-cooling are widely practiced in banana. Banana after packaging in carton should be pre-cooled in pre-cooling chambers till the core temperature of banana attained the storage temperature. It has to be ensured that pre-cooling is done gradually for a period of 4-6 hours to avoid internal damage of the tissue

(Aradhya and Eipeson, 1999). Rodriguez *et al.* (1995) reported that pre-cooling treatment of 5 and 10 minutes combined with storage at 21°C temperature for the first 24 hours or storage at 13°C straight away maintained the green colour of the fruit for 20-22 days. Pre-cooling duration more than 10 minutes were not recommended due to the lowering of pulp temperature close to 13°C, predisposing the fruits to chilling injury.

In export trails of banana conducted by Maharashtra State Agriculture Marketing Board, Pune the pre-cooling was done after packaging the produce in CFB boxes which were moved to the pre-cooling chambers within 10 hours of harvest. The pre-cooling was done at 13°C and 85-90% relative humidity for 8 hours, till the fruit core temperature reached 13°C (Pawar, 1999).

The pre-cooling should be done only if the produce is subsequently stored in cold store or loaded into a reefer container (i.e. only if moving in cold chain). If the produce is going to be subsequently handled or held at ambient temperature, the pre-cooling should be avoided. The fruits after pre-cooling if held at ambient temperature the deteriorative changes will be faster than the one which is not pre-cooled.

As far as possible, environmental conditions (temperature; relative humidity; concentrations of oxygen, carbon dioxide, and ethylene) should be optimized in transport vehicles. Treatment with ethylene to initiate ripening during transportation is feasible, and is commercially used to a limited extent on mature green bananas and tomatoes.

Produce should be cooled prior to loading and should be loaded with an air space between the palletized product and the walls of the transport vehicles in order to facilitate temperature control. Vibration during transportation should be minimized, so as to avoid damage due to bruising. Controlled-atmosphere and precision temperature management should, where possible, be observed so as to allow non-chemical insect control for markets which possess quarantine restrictions against pests endemic to exporting countries and for markets that do not want their produce exposed to chemical fumigants. Mixing several produce items in one load is common and often compromises have to be made in selecting an optimal temperature and atmospheric composition when transporting chilling-sensitive with non-chilling sensitive commodities or ethylene-producing with ethylene-sensitive commodities. In the latter case, ethylene scrubbers can be used to remove ethylene from the circulating air within the vehicle. Several types of insulating pallet covers are available for protecting chilling-sensitive commodities when transported with non-chilling-sensitive commodities at temperatures below their threshold chilling temperatures (Kader and Rolle, 2004).

Chapter – 12

Storage of Banana

The fruits meant for domestic markets are not stored. The consignment meant for long distance transportation is loaded into the refeer vans after pre-cooling and transported . At the destination, the consignment is unloaded and kept in cold storage at 13±1°C. Depending upon the market demand they are ripened using ethylene gas. Normally 75% mature fruits can be stored in green condition upto 4 weeks at low temperature (13±1°C & 90-95% RH). For the local market the dehanded fruits are arranged in plastic crates and sent to ripening chambers for ripening. In traditional system of marketing, the bunches are straight away subjected to smoking upto 12 hours (particularly non-Cavendish varieties) before retailing. But in export trade the dehanded fruits of Cavendish bananas are pre-cooled and stored at 13±1°C with 90-95% relative humidity.

12.1 Shelf Life

Shelf life of banana can be divided into two stages, as green life and yellow life. Most of the banana varieties turn yellow on ripening at ambient temperature, while Cavendish bananas require a ripening temperature of 16-21°C for development of yellow colour. The period for which it remain green is called 'green life' and from turning yellow till un-marketable stage is called 'yellow life'. Cavendish bananas have a longer green life and shorter yellow life, while others have the reverse. The fully mature fruits of Cavendish banana have a shelf life of 6-9 days depending upon the ambient temperature. In local trade, banana fruits are not subjected to pre-cooling or cold storage as it is ripened and consumed immediately. However, in export trade a green life of more than 2 weeks is required to reach the destination market and for its further distribution. Therefore, it undergoes all the post harvest export protocols as detailed elsewhere in this chapter. In order to extend the storage life several methods have been studied by various researchers and a combination of these are used in practice by the traders. The storage life refers to the total life (green and yellow) the fruit including that in storage as well as on the market shelf. While shelf life is the period for which the fruit remains fit for consumption on the market shelf (refrigerated or otherwise). Shaun and Ferris (1997) stated that at ambient tropical temperatures, bananas have an average market life of 1 to 10 days depending on genotype, maturity stage at harvest and storage and handling conditions.

There are several methods of extending the storage life of banana fruits. Harvesting of bunches a few days before they reach full maturity, storage at reduced temperatures (temperature management), modified atmosphere storage, packaging in film bags, pre-treatments with fungicides, chemicals and coating the skin with waxes/emulsions are some of the methods reported. There are several methods of extending the storage life/shelf life of bananas. As was seen in preceeding chapters by adjusting the time of harvest/harvest maturity the shelf life can be managed to some extent.

12.1.1 Maturity Management Vs Shelf life/Storage Life

Banana being a climacteric fruit, ripens naturally after attaining the full maturity. On attaining full maturity or due to abiotic or biotic stresses the fruit shifts into senescence phase, associated with a rise in rate of respiration and ethylene evolution. Fully mature fruit responds vigorously to the auto-catalytic property of endogenously evolved ethylene. While the response of less mature fruits to ethylene is dose dependent. Therefore, lesser mature fruits have longer green life compared to full mature. Fruits with 75% maturity can be stored upto 3-4 weeks in green condition at low temperature (13$\pm$1°C; 90% RH), while the more mature fruits have relatively lower storage life. However, taking the yield and quality into consideration, harvesting banana fruits with less than 75% maturity is not advisable as it results in lesser yields. Studies using Cavendish banana fruits harvested from 65 to 105 days after flowering showed that after the onset of climacteric, little differences in ripening characteristics were found among the fruits harvested at different maturities. When bananas are used for processing, commercially acceptable pulp quality is attained only after a minimum of 80 days after shooting. Studies have revealed that when both early and late harvested Gaint Grovernor (AAA) bananas were ripened and compared for its quality, late harvested fruits showed higher percentage of total soluble solids, reducing sugars and total sugars and lesser starch and acidity. Poovan (Mysore, AAB) and Karpuravalli (Pisang Awak, ABB) varieties of banana also showed similar results (Narayana and Mustaffa, 2007). However, lesser mature fruits had more weight loss during storage compared to full mature fruits.

In Robusta banana, by harvesting the fruits at early stage of maturity (at 100 days) the shelf life in ambient conditions could be extended from 16-21 days (Shantha Krishnamurthy and Krishnamurthy, 1989), while cv. Poovan when harvested at 75% maturity, de-handed and either kept at room temperature (23-28°C, control) or covered with moist sand (20-26°C) could be stored upto 30 days. Control fruits ripened in 10-13 days (Poovan) and 13-15 days (Robusta) compared to 20-21 and 22-30 days respectively for those kept in sand.

12.1.2 Temperature Management Vs Shelf life/Storage Life

Optimum temperatures for storage and holding of green bananas are 13.3 to 14.4 °C (56 to 58 °F) (Thompson and Burden, 1995). Some banana cultivars can be handled at 12.8 °C (55 °F) too. Optimum RH for holding and transport of fruit is 90 to 95%. Ripe bananas cannot be held for long as compared to green. Storage temperature influences the post harvest physiology of banana and its interaction with the microbes in the storage environment. As the ripening is hastened at higher temperatures, the complex carbohydrates like starch degrade into simple sugars to sustain the high rate of respiration. Further the self defense of fruit weakens, and the microbes in the storage environment attack the fruits resulting in spoilage and rots. Shelf life was found to decrease exponentially with increasing temperature (Peacock,1980). A reduction in acid-phosphotase activity and failure to ripen were noticed when mature green bananas were stored at a constant 40°C in contrast with controls stored at 20°C.

Lowering of temperature slows down the catabolic process in the fruit and also makes the environment non-congenial for the growth and proliferation of spoilage microorganisms. Storage of bananas at 13°C considerably suppressed respiration rate, ethylene production and ripening. Normally bananas are stored at a critical temperature of 13$\pm$1°C with a relative humidity of 85-95%. Below this temperature fruits suffer chilling injury. However, variation in varietal response to chilling temperatures and manifestation of chilling injury symptoms have also been reported.

The produce for overseas markets are carried in refrigerated ships or 'reefer ships', which maintain a circulating cool air temperature of 12-13°C. Under a normal low temperature storage condition (14.5$\pm$1°C; 80-90% RH), after 4 weeks, 64% of Robusta banana were found to be fully ripe and 100% of Dwarf Cavendish overripe. Therefore, ideally it can be kept green only for 3 weeks.

Green unripe fruits at 2 stages of maturity (100 and 115 days after fruit set) when held at room temperature or at 15 and 20°C for 1-4 weeks followed by ripening at ambient conditions showed that the quality of fruits after 1-4 weeks of holding at 20°C was the best in quality. As the storage duration at low temperature prolongs, the post storage ripening quality decreases.

12.2 Pre-Treatments Vs. Shelf Life/Storage (Hot Water, Fungicides, Coatings and other Chemicals)

12.2.1 Hot Water Treatment

Pre-treatments with hot water, fungicides, wax emulsions have also shown beneficial effect in extending the storage life at ambient as well as low temperature. Hot water treatment is an effective non-chemical method of postharvest pest and disease control if combinations of suitable temperatures and exposure times are selected to prevent the quality loss of produce (Lurie, 1998). However, hot water treatment is limited to a narrow range of temperatures and time intervals and those values are highly dependent on the commodity. Furthermore, efficiency of hot water treatment could vary with crop variety, agronomic practices employed in the field and climatic zones where the crops are grown. Universally effective temperatures and exposure times to all postharvest pathogens are not available (Barkai-Golan and Phillips, 1991). Optimum temperature and exposure time for postharvest hot water treatment of banana were determined to be 50°C and 3 min, respectively for Embon (AAA type) and Kolikuttu (AAB type) varieties of banana in Sri Lanka. Temperatures higher than 50°C reduced the colour of fruit peel and gave a pale appearance to fruit even after ripening. Exposure times longer than 5 min substantially reduced the Brix value (De Costa and Erabaduputiya, 2005). The hot water treatment at 51°C for a period of 15-20 minutes extended the storage life of Neypoovan banana up to 30 days at 13.5°C without significantly affecting the quality of fruits. Similarly it was found that hot water treatment at 46°C for 30 minutes was best for extending the storage life of Rasthali banana up to 32 days at 13.5°C (Narayana, 2003).

12.2.2 Fungicidal Treatment

Anthracnose, crown rot and blossom end rot are common and serious postharvest diseases of banana. Anthracnose, caused by the fungus *Colletotrichum musae*, is a latent infection. The fungal spores infect immature banana in the field but symptoms appear only at the ripening stage. A range of fungi, including *C. musae* causes crown rot as an infection developing on the cut surface of banana hands. Blossom end rot starts from the tip of the finger during ripening and is caused by fungi complex including *C. musae*. Physiological and storage conditions at the postharvest stage, accelerate the development of the above diseases. Wound anthracnose, caused by *Colletotrichum musae*, and early ripening is the main problem affecting the quality of export bananas (*Musa* spp.) from a lot of countries in the world. Sarkar *et al.* (1995) reported that Giant Governor bananas could be kept for 14 days after harvest without significant effects on their quality if they were treated with GA_3, Blitox, Bavistin, Dithane M-45 or 6% waxol.

In Guadeloupe (French West Indies), these problems generally concern bananas harvested in lowland plantations during the rainy season (Chillet *et al.*, 2010). The efficacy of benomyl, prochloraz, flusilazole and other fungicides in controlling crown rot, was studied by Jones (1991). In experiments with harvested fruit, prochloraz and flusilazole were the most effective fungicides compared to benomyl.

Until the early 1980s, banana hands were dip-treated against crown rot with suspensions of benomyl and/or thiabendazole (TBZ) and 0.5% alum (aluminium sulphate) in banana packing plants. In 1982, the dip treatment was replaced by the use of fungicide-impregnated cellulose pads which were placed on the cut surface of the crown. Originally, the pads were impregnated with TBZ and alum. From 1990 onwards, a combination of TBZ, imazalil and alum was used. In the pad system, hands were severed from the rachis in the field and placed on a green banana leaf with the crown facing downwards for a recommended period of 2 min to allow excess latex to drain out. The person harvesting would place one pad firmly onto each crown before packing in boxes in situ.

Crown pads were phased out by the introduction of the mini wet-pack (MWP) system in 1993 which is still in use today. In the MWP, bananas are dip-treated in a fungicide suspension and packed while still wet. But in contrast to Latin American banana post-harvest facilities, the operation is decentralised to suit smallholders (Krauss and Johanson, 2000). The procedure is as follows.

Crown rot control and packing are carried out in small sheds constructed by the farmers at their fields. Fruits are delatexed in the field, usually after the hands have been cut into smaller clusters (4-8 fingers) ('clustering'). Cutting the clusters in the shed would be preferred (Krauss and Operation 2000/Geest, 1996a) to avoid wound infection during handling prior to the dip treatment. The recommended delatexing time is 10 minutes. Even with the MWP system, adherence to the recommended time for delatexing is erratic and is sometimes substantially exceeded, especially when the cutting operation takes place at a distance more than 10 min walk away (round-trip) from the packing shed. In the shed, clusters are submerged in a fungicide suspension in tubs of 22.5 or

45 litres for a recommended time of 10 seconds. TBZ (300 ppm) or imazalil (500 ppm) as Fungaflor 75SP' (Jannsen Pharmaceuticals) both in conjunction with alum (0.5% w/v) are recommended.

Of the seven fungicides registered for use in banana industry, only two are designated for post-harvest treatment: thiabendazole (TBZ) as "Mertect 20-S" and imazalil as "Fungaflor 75SP". Several alternative fungicides have been evaluated for crown rot control in other banana growing countries but these were largely on an experimental scale.

Dipping of bananas in 200 ppm thiabendazole was an approved and recommended post harvest treatment for export of bananas. When effect of various pre-treatments like dipping in thiabendazole at 400 ppm or 5% Quinolate, wrapping in PE bags before packing in cartons were evaluated for banana cv.Magrabi, it was found that all packaged fruits were in good condition after one month of storage and had a shelf life of 5-7 days in ambient conditions. Treatments with GA_3, Blitox, Bavistin, Dithane M-45 or 6% waxol have also shown beneficial effect in extension of storage life of Giant Governor bananas.

Although chemical control is a feasible option to control postharvest diseases, it has high environmental and health risks (Eckert and Ogawa, 1985; Ragsdale and Sisler, 1994; Mari and Guizzardi, 1998; Janisiewicz and Korsten, 2002). Therefore, alternative non-chemical methods available for control of postharvest diseases are being explored.

12.2.3 Natural Biocontrol Agents

Consumer demand is increasing for fruit which has been treated non-chemically for postharvest pathogens. Naturally occurring antagonists on host surfaces are a promising component of biological crop protection. Success of different natural antagonists such as fungi, bacteria and yeast, especially for the control of wound-invading, secondary pathogens of fruit has been reported (Mari and Guizzardi, 1998; Janisiewicz and Korsten, 2002). Effectiveness of biological control as the sole method of controlling postharvest pathogens is reported to be low when compared to that of synthetic fungicides (Droby *et al.*, 1993, 1998; El-Ghaouth and Wilson, 1995). Efficacy of biological control can be improved by manipulating the environment and integrating with other methods of control (Janisiewicz and Korsten, 2002). Integrated methods of biological control with fungicides as field sprays have shown effective control of anthracnose of mango and avocado (Korsten *et al.*, 1997). A member of the *Burkholderia cepacia* complex, isolated from the fructosphere of banana was found to be effective as an antagonist of postharvest pathogens even after 5 years of storage in sterile distilled water at ambient temperature. The most effective concentration of *B. cepacia* was determined to be 10^{10} CFU/ml for in-vivo control of anthracnose and crown rot. The most effective control was achieved by repeated dipping of bananas in bacterial solution or by adding a wetter (Tween 20) to the bacterial solution (De Costa and Erabaduputiya, 2005). Combining hot water treatment with the bacterial antagonist also gave more effective control of anthracnose, crown rot and blossom end rot than using the two treatments individually. Higher efficacy of this integrated method was confirmed further by its greater control of artificially infected anthracnose.

12.2.4 Mineral Oils

Oils had been used as pesticides for centuries and are some of the most effective safe alternatives to synthetic insecticides and fungicides. Mineral oils are consider a promising control agent against a wide varieties of pests all over the world. Petroleum oils are highly refined, parafinic oils that are used to manage pests and diseases of plants. Oil sprays work by smothering the insects and mites they come in contact with, preventing several disease causing pests that make a home out of favored plants like roses, hydrangeas, phlox, and zinnias. When oil is mixed with other ingredients it can also prevent or control fungal diseases.

Petroleum oils may be referred to many names, including horticultural oil, spray oil, dormant oil, summer oil, supreme oil, superior oil, Volck oil or white mineral oil. Mineral oils are a type of refined petroleum product, and horticultural oils (Sunspray, Scalecide, Volck) are mineral oils that have been distilled to remove impurities that can damage plants. The final formulations of these oils are combined with an emulsifying agent that allows the oil to mix with water. The mode of action of these oils is said to be due to the reduction of spore production by smothering the infected surface (Northover and Schneider,1996).

Post-harvest antifungal treatment with an emulsifiable mineral oil (E 9267 oil) was compared with thiabendazole, carbendazim, clorothalonil, thiofaminet and Aureofungin for the control of storage decay of Robusta banana. Pre-storage dip in the E 9267 oil (0.4%) was found very effective in reducing the storage decay caused by anthracnose and prolonged the storage life (Seshadri *et al.*, 1980).

12.3 Surface Coatings

Edible coatings have long been known to protect perishable food products from deterioration by retarding dehydration, suppressing respiration, improving textural quality, helping retain volatile flavor compounds, and reducing microbial growth (Debeaufort *et al.,* 1998). Edible coatings are applied on fruits and vegetables to improve appearance, delay ripening, reduce water loss and decay, and extend shelf life, but may also change flavour (Baldwin *et al.*, 1995; Saucedo-Pompa *et al.*, 2007; Hoda *et al.*, 2012). Edible coatings create a modified atmosphere within the fruits and delay their ripening process. Type and concentration of edible coating components contribute in determining the quality (changes in physicochemical characteristics) of coated bananas. Due to the hydrophobic nature of banana cuticle, addition of emulsifiers such as sodium caseinate can enhance the homogeneity of the edible coating on the banana skin. Edible coatings also work as a replacement for natural protective waxy coatings and act as physical barrier towards carbon dioxide, oxygen and moisture movement for the fruits (Baldwin *et al.*, 1999). Protein and polysaccharide coatings have excellent barrier properties against O_2 and CO_2. Small amount of additive such as plastisizers like glycerol can improve the appearance of coated fruit and mechanical property of the edible coating. Therefore, composite edible coatings can successfully be employed as a means to improve the barrier characteristics of edible coatings covering the banana and extend its shelf life.

Coatings can be formulated from different components such as hydrocolloids (polysaccharides and proteins), lipids (waxes and resins) and synthetic polymers. These edible materials have different barrier properties against gases and physico-chemical and mechanical characteristics. Therefore, most coatings are made of more than one material with the addition of low molecular weight molecules including sorbitol, polyols or glycerol that serves as plasticizers (Risse and Miller, 1983; Olivas and Barbosa-Canovas, 2005). Semi-permeable coatings can also modify internal atmosphere of fruits by changing the composition or concentration of standard atmosphere gases like CA storage, with less expense incurred (Nisperos-Carriedo *et al.*, 1992). The atmosphere created by coatings is strongly related to permeability of coating and fruit respiration rate. (Baldwin *et al.*, 1996). Coating Giant Cavendish banana fruits with Pro-long 24 hours after initiation of ripening decreased their skin permeability to gases and depressed their oxygen content. It also suppressed rapid ethylene production and climacteric rise. Coating delayed the chlorophyll loss, which normally accompanies ripening (Banks, 1985).

'Hindi' and 'Paradica' varieties of bananas coated with 1.5% Prolong (sucrose ester) delayed the ripening significantly (Hussein *et al*, 1985). Fruit coated with polyethylene parafilm wax at concentrations of 20%, 25%, and 30% (v/v) and then held at 29–30°C for 5 days delayed early peel spotting compared with controls. This delay in peel spotting, after surface coating, is believed to be due to reduced PAL activity. Low rate of oxygen diffusion through the coating might be the factor that limits the last step to blackening (Promyou *et al*, 2007). A double coating of 12% wax emulsion was found to prolong the storage life of Dwarf Cavendish bananas by 10-12 days at 14.5°C. Kolekar *et al* (1988) observed that the surface coating of bananas with sucrose stearate emulsion (1.0%) delayed ripening of the fruit for 4 days, while Islam & Kabir (2001) observed that wax coating of banana fruit increased the shelf life to 21 days. Treatment with reduced oxygen for short periods (1– 3 days) before storage has also been found to retard the pre-climacteric ripening of bananas by delaying the climacteric peak (Pesis *et al.*, 2001).

Green unripe fruits at 2 stages of maturity (100 and 115 days after fruit set) when held at room temperature or at 15 and 20°C for 1-4 weeks followed by ripening at ambient conditions showed that the quality of fruits after 1-4 weeks of holding at 20°C was best. By harvesting the fruits early (at 100 days) the shelf life in ambient conditions could be extended from 16-21 days (Santha Krishnamurthy and Krishnamurthy,1989). Green bananas coated with the quick dip method in a sucrose ester formulation (0.5 – 2%) were stored with non-treated fruits at ambient temperature for 10 days. The treatment delayed ripening and the best results were obtained with 1% emulsion (Kolekar *et al.*, 1988).

Banana cv. Poovan and Robusta were harvested at 75% maturity and the bunches were cut into hands which were either kept at room temperature (23-28°C, control) or covered with moist sand (20-26°C) for up to 30 days. Control fruits ripened in 10-13 days and 13-15 days for each cultivar respectively compared to 20-21 and 22-30 days for those kept in sand (Neelakantan and Sadasivan, 1977). Ag (I) is reported to retard ripening of banana at 800-ppm concentration and extended storage life up to 28 days at 15°C (Rouhani, 1978). As far as the quality changes are concerned highly significant periodical variations in contents of total sugars, starch, titrable acidity, ascorbic acid, total chlorophyll and yellow pigments were observed in fruits of the varieties Pachabale, Rasabale and Rajabale during storage at 20°C. Increase in the mean pulp to skin ratio of 3 varieties of banana fruits (from 1.77 to 2.91), total reducing sugars (0.18 to 14.22%), titrable acidity (2.35 to 6.14 mg/100g pulp) and ascorbic acid (2.40 to 7.08 mg/100g pulp) was observed during the storage of banana fruits at 20°C.

Recently a renewed interest is being shown on use of chitosan in post harvest treatments of fruits and vegetables. It is a natural product obtained from crustaceans shells. Chitosan, is a cationic polysaccharide with high molecular weight, obtained by the alkaline deacetylation of chitin (a long-chain polymer of a *N*-acetylglucosamine) and is soluble in dilute organic acids (El Ghaouth *et al.*, 1991; El Ghaouth *et al.*, 1992; Cheah *et al.*, 1997). Sensory evaluation indicated that coated longan fruit with chitosan was better in quality when compared to controls (Jiang and Li, 2001). The chitosan concentration appeared to be the most significant ($P< 0.1$) factor influencing all variables except for TSS. The optimum concentration of chitosan and glycerol were predicted to be 2.02% and 0.18%, respectively (Hoda *et al.*, 2012).

Doshi and Sutar (2010) evaluated the effects of chitosan 1% and 1.5%, calcium chloride ($CaCl_2$) 1% and 1.5%, chitosan 1% + gibberellic acid 100 ppm, chitosan 1.5% + gibberellic acid 100 ppm, jojoba wax, and glycerol (98%) coatings on the shelf life and postharvest quality characteristics of banana fruits stored at 34 ± 1°C and 70–75% relative humidity. The coatings of chitosan, chitosan + gibberellic acid, and jojoba wax delayed the changes in the weight loss percentage, decay percentage, total soluble solids, pH, titrable acidity, sugar accumulation, pigment degradation, and ascorbic acid compared to uncoated ones. Further, the least disease incidence was found to occur in the banana fruits treated with chitosan and chitosan + gibberellic acid. They recommended that coating with chitosan and chitosan + gibberellic acid has the potential to control decay percentage, prolong the shelf life, and preserve valuable attributes of banana. Calcium chloride and glycerol coatings neither showed any potential to control disease incidence nor prolong the storage life and preserve valuable attributes of banana fruit.

Sodium carboxymethyl cellulose, sodium caseinate and glycerol are the components used as plasticizers in edible coatings. The optimum concentrations of sodium carboxymethyl cellulose, sodium caseinate and glycerol for coating banana were predicted to be 1.32, 0.40 and 0.86%, respectively. In general, the sodium caseinate content appear to be the most significant factor influencing all characteristics of coated banana studied at 26±2C and 40–50% relative humidity (Hoda *et al.*, 2012.)

12.4 1-MCP Gas Exposure

In recent years, some very effective compounds that block production, action, synthesis or compete for its binding sites, such as 1-methylcyclopropenee (1-MCP), are being used for extending the shelf life of fruits, vegetables and ornamentals. 1-Methylcyclopropene (1-MCP) is a cyclopropane derivative used as a synthetic plant growth regulator. The mechanism of action of 1-MCP involves its tightly binding to the ethylene receptor in plants, thereby blocking the effects of ethylene. Jiang *et al.* (1999) reported that the application of 1-methylcyclopropene (1-MCP) in combination with polyethylene bags extended the postharvest life of banana fruit. It has been shown that the effectiveness of 1-MCP was inconsistent when applied to ethylene-pretreated bananas (Pelayo *et al.*, 2003). Most of these studies have focused on practical applications of 1-MCP to extend postharvest life when it is applied to fruits at the pre-climacteric stage, that is, before the induction of autocatalytic ethylene. Studies on extension of the shelf life of post-climacteric (ripe) fruits are scanty. 1-MCP combined with perforated polyethylene packaging extended shelf life of banana for over two weeks without much loss in marketability (Zewter *et al.*, 2012).

12.5 Use of Botanicals

Plant extracts for the control of plant disease are emerging as alternatives to conventional fungicides as they are generally safe to humans and environmentally friendly. The extracts of 50 plants were screened *in vitro* against the anthracnose causing fungal pathogen *Colletotrichum musae*. The extracts of *Solanum torvum*, *Jatropha curcas* and *Emblica officinalis* (*Phyllanthus emblica*) inhibited the mycelial growth of *C. musae*. *S. torvum* extract at 25 and 50% (w/v) completely inhibited the growth of *C. musae* while those of *E. officinalis* and *J. glandulifera* restricted growth to 7.6 mm day^{-1} at 50% concentration. The same extracts were tested *in vivo* at room temperature (28±2°C) and in cold storage (13.5°C) against anthracnose disease on the banana cultivars Robusta (AAA), Rasthali (AAB) and Ney Poovan (AB), *S. torvum* extract was very effective in reducing the incidence of the disease, better than a standard treatment with the fungicide benomyl (0.1%). The extracts also significantly increased the shelf life of bananas, particularly their green life, where *S. torvum* extract was the best. The shelf life was extended by 16-20 days compared to the control; in most cases the extension was significantly longer than that afforded by benomyl. Hence, plant extracts, particularly of *S. torvum*, can be used not only for the management of anthracnose disease but for increasing the shelf life of bananas (Thangavelu *et al.*, 2004).

Mechanism of disease suppression by plant products have proved that, the active principles present in them and/or induce systemic resistance in host plants. In an investigation by Jadesha *et al.* (2012) the expression of induced systemic resistance in banana fruits upon treatment with plant extracts was seen. Banana fruits dipped in 10 per cent of leaf extracts of *Solanum torvum*, Zimmu and *Allium alliaceum* for five minutes and inoculated with conidial suspensions (106/ml) of *Colletotrichum musae* by pin prick method were analysed for PO, PPO and PAL enzymes separately in peel and pulp 2, 4, 6, 8 and 10 days after treatment. The Peroxidase(PO), Polyphenol oxidase(PPO) and Phenylalanine ammonia-lyase(PAL) activity was significantly increased both in peel and pulp of the inoculated fruits dipped in leaf extracts as compared to inoculated fruits alone. Activity of defense enzymes was found to be highest on sixth day after inoculation and decline in subsequent days.

Botryodiplodia theobromae,. Aspergillus niger and *Aspergillus flavus* were isolated from crown rot affected 'Medium Cavendish' banana. When the *Botryodiplodia theobromae* inoculated bananas were treated with extracts of *Moringa oleifera* (leaf extract), *Azadirachta indica* (seed extract) and *Cassia alata* (leaf extract) and *Zingiber officinale* (rhizome extract) it was found that at a concentration of 66.67% w/v *Zingiber officinale* (rhizome extract) was the most effective against crown rot disease (Kumah *et al.*, 2010).

There is abundant scientific literature on antibacterial (Cellini *et al*, 1996), antiparasitic (Lun *et al.*, 1994), antifungal (Davis *et al*, 1990) and antiviral (Weber *et al*, 1992) activities of garlic. Sodium metabisulphite has been used as an antibrowning agent on bean, apples, potatoes and banana (Brecht *et al.*, 1992). McFeeters *et al.* (2004) used sodium metabisulphite to prevent darkening and as a microbial inhibitor in acidified vegetables and found to prevent rapid softening reaction. The treatment of fruit with 1% garlic extract or 10% onion bulb extract and adjusting the pH to 4.5 extended the shelf life by 4–5 days when compared with untreated fruit. Treatment with a combination of 1% garlic extract and 1 mm sodium metabisulphite and adjusting the pH to 4.5 prolonged the shelf life of ripe banana to 13–14 days when compared with 6–7 days for the untreated fruit. The treatment delayed the fruit softening by diminishing the rate of starch and pectin degradation. The rate of increase in β-amylase and polygalacturonase activities along with softness was also retarded in treated fruits when compared with control fruits (Sanwal and Payasi, 2007).

12.6 Other Chemicals/Methods

12.6.1 Wax Emulsion Coating

When Taijiao bananas were harvested at 70-80% maturity and sprayed with Jiaoxuan-1 or Jiaoxuan-2 (preservative agents) or with carbendazim (fungicide), wrapped in PVC film and stored at 11-25°C, they could be stored up to 90 days with least weight loss (Huang and Chen, 1993). Green bananas coated with sucrose ester formulation (0.5 – 2%) in the quick dip method delayed the ripening upto 10 days under ambient conditions.

12.6.2 Ethylene Absorbent

Preventing build up of ethylene around produce is among the methods in use to delay ripening of bananas. This is achieved through the use of agents that absorb ethylene such as $KMnO_4$ (Shaun and Ferris, 1997) or block ethylene binding to its receptor (Sisler *et al.*, 1999). Use of ethylene scrubbers were found to have complimentary effect in extending the shelf life and storage life of banana when used in polybags or storage chamber. Potassium permanganate is an effective ethylene absorbent and several commercial formulations of ethylene scrubbers based on this are available in the market. Banana fruit stored with ethylene absorbent at 30°C was reported to have a similar storage life to that of conventionally stored fruits in air at 12.8°C. This is a very useful technique especially for developing countries where refrigeration is not readily available or is expensive (Scott and Soertini, 1974). They demonstrated the use of vermiculite block soaked in potassium permanganate solution to remove ethylene from sealed bags containing banana. Purafil (a silicate carrier impregnated with alkaline $KMnO_4$) is a commercial formulation available for this purpose, which extended the storage life of pre-climacteric bananas upto 16 days at 13°C and 8 days at 24°C when kept in sealed polyethylene bags. While various materials like vermiculite, coke, alumina, lime stone, pumice stone, sand, brick pieces and sawdust were used as a carrier for $KMnO_4$ based ethylene absorbent, use of other chemicals like charcoal and palladium chloride was also found to prevent the accumulation of ethylene in banana packs.

12.6.3 Salicylic Acid

Salicylic acid, an ubiquitous plant phenolic, has been reported to regulate a number of processes in plants (Raskin, 1992a). One of its earliest known *in-vivo* functions in plants is heat production during flower induction in some angiosperm species (Raskin *et al.*, 1987). Later it was shown to regulate expression of pathogenesis related protein genes, which suggested its role as a signal molecule in providing resistance against pathogen attack. Salicylic acid mediated hypersensitive and systemic acquired resistance against pathogen attack are proposed to be mediated through the inhibition of catalase, which subsequently raises intracellular H_2O_2 concentration. This increased intracellular H_2O_2 concentration has been proposed to act as a second messenger in activation and expression of defense related genes (Raskin, 1992a). Recently, it has been proposed to be a new kind of plant hormone (Raskin, 1992b). Thus, salicylic acid has been shown to interfere with the biosynthesis and: or action of ethylene (Raskin, 1992a), abscisic acid (Apte and Laloraya, 1982) and cytokinins (Sano *et al.*, 1994) in plants. Salicylic acid and its derivative acetyl salicylic acid (ASA) have been reported to inhibit ethylene production in pear (Leslie and Romani, 1986 & 1988).

Salicylic acid treatment has been found to delay the ripening of banana fruits (*Musa acuminata*). Fruit softening, pulp:peel ratio, reducing sugar content, invertase and respiration rate have been found to decrease in salicylic acid treated fruits as compared with control ones. The activities of major cell wall degrading enzymes, viz. cellulase, polygalacturonase and xylanase were found to be decreased in presence of salicylic acid. The major enzymatic antioxidants namely, catalase and peroxidase, were also found to be decreased in presence of salicylic acid during banana fruit ripening (Srivastava and Dwivedi, 2000).

12.7 Evaporative Cool Storage

The basic principle depends on cooling by evaporation. As water evaporates it has a considerable cooling effect, and the faster the rate of evaporation the greater the cooling. Evaporative cooling occurs when air, which is not already saturated with water, passes over a wet surface. An evaporative cooler thus consists of a wet, porous bed through which air is drawn. The water evaporates into the air raising its humidity and at the same time cooling the bed. The efficiency of an evaporative cooler depends on the humidity of the surrounding air. Very dry, low humidity air can absorb a lot of moisture so considerable cooling occurs. In the extreme case of air that is saturated no evaporation can take place and no cooling occurs. Weight loss is one of the common changes observed in fresh produce during storage and ripening due to transpiration and respiratory loss of carbohydrates. Cool chamber storage significantly slowed down the weight loss (PLW) and decrease in firmness of banana Cv. Dwarf Cavendish fruits compared to rapid changes in fruits stored in ambient conditions. The banana stored in the evaporative cool chamber developed by Prof.S.K.Roy exhibited a shelf life of 20 days with only 2.50% weight loss compared to those stored at ambient temperature which had only a shelf life of 14 days with a weight loss of 4.80%.

The shelf life and qualities of banana in the perforated plastic container were much longer and higher than those inside the sisal sack and control. Also, the packaging of produce inside the perforated plastic container retained the highest (13.1%) amount of vitamin C content (Alabadan *et al.*, 2008). Differently pre-cooled banana fruits when stored in evaporative cool chamber (25 ± 1°C at 90 - 95% RH) showed an enhanced shelf life of 3 days compared to ambient temperature (30 ± 3°C) (Doshi and Sutar, 2010). Waskar and Roy (1993) reported that storage of banana in zero energy cool chamber is advantageous over cool store as it does not require any electrical energy and is easily installable.

Cool chamber storage significantly slowed down the rate of increase in PLW and decrease in firmness of banana Cv. Dwarf Cavendish fruits compared with rapid changes in fruits stored in ambient conditions (Nagaraju and Reddy, 1995).

In an experiment conducted to see the effect of packaging (400 gauge LDPE bag) with and without ventilation on shelf-life and quality of Karpuravalli banana at ambient, zero energy cool chamber and 13.5°C, it was found that, the shelf-life in unvented polybags was maximum (19.33 days) at 13°C followed by zero energy cool chamber (7.33 days) and the least was under ambient conditions (6.67 days) as compared to 19, 8.33 and 7.67 days respectively with ventilation. Among the quality parameters, non-significant variation was observed for TSS of fruits under all the three storage conditions.. The acidity increased gradually up to the end of green life and then decreased (in unvented polybags) or remained same (in ventilated and control) till the end of yellow life. Reducing sugars and total sugars increased gradually throughout the storage period and was maximum in control and ventilated bags than the unvented polybags. Spoilage was maximum in unvented polybags especially under ambient and zero energy cool chamber conditions. Storing of Karpuravalli banana fruits in unvented polybags at low temperature could significantly increase the green life up to 19.33 days while storage under zero energy cool chamber or ambient condition requires ventilation (0.5%) to reduce spoilage (Narayana *et al.,* 2002).

12.8 Modified Atmosphere Storage (MAS)

Modified atmosphere storage is achieved by using polymeric films, sometimes using selectively permeable films which have a specific rate of exchange for oxygen, carbon dioxide and water vapour. However, in this system there is no attempt to control the atmospheres of oxygen, carbon dioxide and nitrogen at specific concentrations unlike control atmosphere storage. Use of modified atmosphere and vacuum packaging has also been found to retard the ripening of ethylene-pretreated banana (Pesis *et al.*, 2005).

Storage life of banana can be extended by packaging in sealed polyethylene bags with ethylene absorbent (Satyan *et al.*, 1992; Scott and Soertini, 1974). The objective of using polyethylene bags is to maintain high humidity and create a modified atmosphere around the produce in the package. Though banana release a remarkable level of ethylene, inclusion of an ethylene-absorbing compound, such as potassium permanganate on an inert carrier, in the package can reduce accumulation of ethylene considerably. A simple polyethylene bagging was adequate for extending the storage life by about a week at warm ambient temperatures, while longer storage could cause ripening of green bananas due to accumulation of ethylene gas. Satyan *et al.* (1992) observed that when the whole banana bunches were packed in sealed polyethylene tubes and held at 29°, 20° and 13°C, the average storage life was increased 2- to 3-folds (over control). Further increase of 3- to 4-folds was observed when an ethylene absorbent ($KMnO_4$ on vermiculite) was packed with the bunch in PE tubes. Similar findings were also reported in Lacatan banana (Rao and Chundawat, 1991). The storage life of bananas (Dwarf Cavendish and Williams) was extended when fruit were held in a modified atmosphere containing 10% carbon dioxide for 1 to 3 days, followed by storage under regular atmosphere conditions at 12.5°C.

Studies were conducted by Sarkar *et al.* (1997) to determine the suitable thickness and colour of polyethylene films used to extend the shelf life of banana cv. Giant Governor. They found that physiological loss in weight (PLW) was least for fruits packed in 300 or 400 gauge, non-perforated polyethylene packs and fruits remained in marketable condition up to 28 days after harvest.

Effect of pre-treatments (dipping in thiabendazole at 400 ppm or 5% Quinolate), storage temperature (13 or 15°C) and packaging (wrapping in PE before packing in cartons) on storage quality of banana cv.Magrabi were investigated by Elazayat (1996). All packaged fruits were in good condition after one month of storage and had a shelf life of 5-7 days in ambient conditions. These fruits ripened normally following storage. Control fruits were distorted in shape and decayed after storage.

When Taijiao bananas were harvested at 70-80% maturity and sprayed with Jiaoxuan-1 or Jiaoxuan-2 (preservative agents) or with carbendazim, wrapped in PVC film and stored at 11-25°C, they could be stored up to 90 days (Huang and Chen, 1993).

12.9 Controlled Atmosphere Storage (CAS)

Commonly CAS is the term used for a high carbon dioxide, decreased oxygen and high nitrogen levels as compared to normal atmosphere (Mohapatra *et al.*, 2010). Controlled atmosphere storage is a system for holding produce in an atmosphere that differs substantially from normal air in respect to CO_2 and O_2 levels. Unlike MAP, the Controlled atmosphere storage requires the constant monitoring and adjustment of the CO_2 and O_2 levels within gas tight stores or containers. The gas mixture will constantly change due to metabolic activity of the respiring fruits in the store and leakage of gases through doors and walls. The gases are therefore measured periodically and adjusted to the predetermined level by the introduction of fresh air or nitrogen or passing the air in store atmosphere through a chemical to remove CO_2.The concentration of carbon dioxide, oxygen and nitrogen are manipulated to attain various gas compositions as desired depending on the commodity. CA storage also have adverse effects on the produce, at O_2 levels below 1%. In the absence of O_2, anaerobic conditions can prevail with the consequent formation of alcohol, aldehydes and other physiological changes. Also high CO_2 and low O_2 may cause abnormality in metabolism. Bananas were found to suffer carbon dioxide injury at more than 7% concentration and low oxygen injury at levels less than 1%. It was observed that reduced O_2 and increased CO_2 inhibited the action of ethylene such that the ripening processes of ethylene treated bananas were delayed.

For CA storage of Dwarf Cavendish and Williams bananas 2% oxygen and 7% carbon dioxide was ideal to keep the fruits in green condition for six weeks at 12.5°C. Storage of bananas in CA shipping containers or packing in polyethylene bags and keeping in standard shipping containers was also successful for export of bananas by sea (Paunescu and Bratucu, 2010). Controlled atmosphere (CA) storage composition of 10 % CO_2 and 5 % O_2 was found to be optimum at RT while 5 % O_2 and 5 % CO_2 was optimum at 15°C storage with extended storage life and quality maintenance for Robusta banana. Robusta banana could be kept in unripe condition for 25 days at RT, 56 days at 15°C and 75 days at 13°C which ripened normally within a week when shifted to ambient conditions

12.10 Vacuum Packaging

Vacuum packaging is a special type of modified atmosphere packaging, wherein a partial vacuum is created in the head space, leaving an altered initial atmosphere. By vacuum packaging (in polyethylene bags), the storage life of Golden bananas was reported to be extended upto six weeks under refrigerated condition (17°C) and upto three weeks at room temperature. Though there are few reports on usefulness of vacuum packaging in extending the shelf life of bananas, studies at NRC for Banana, Trichy showed that it is not suitable for long term storage of green bananas, as the fruits failed to ripen after removal from vacuum due to damage of the ripening apparatus of fruit beyond recovery (Anon., 2009a). The fruits also suffered carbon dioxide injury characterized by surface blackening and fermented off flavour. Partial vacuum may have beneficial effect when coupled with low temperature storage. The normally available films are not meant to maintain the absolute vacuum, as they are permeable to air and water vapour. Therefore, over a period of time such packs pull air from atmosphere due to negative pressure inside the pack and lose vacuum. Later due to buildup of carbon dioxide inside the pack, the fruits suffer carbon dioxide injury.

12.11 Hypobaric Storage

Hypobaric storage, also known as low pressure storage, is a type of CA storage with emphasis on reducing the pressure exerted on the stored material. This method reduces oxygen concentration in the storage environment and also evacuates the ethylene in the tissues by diffusion. Produce is stored in a partial vacuum and chamber is vented continuously with water-saturated air to maintain oxygen levels and minimize water loss. The reduced partial pressure of oxygen in hypobaric storage retards ethylene action and its production as well as ripening in some fruits. Extended storage life of 120 days have been reported when stored at a partial pressure of 150 or 80 mm Hg. The rate of ripening was found to be inversely related to the pressure during storage. When Dwarf Cavendish bananas were pre-treated with ethylene and stored for 28 days in 1/10th atmospheric pressure at 14°C, it remained green and firm until the end of storage, but started ripening almost immediately after being placed in 21°C without additional ethylene treatment. This indicates that sub-atmospheric storage (hypobaric storage) retains the produce in fresh condition much longer than conventional cold storage.

12.12 Irradiation

Irradiation, also known as radurisation, is the process of exposure of fruits to ionizing radiations. Though doubts have been raised on the prospects of using ionizing radiation for extending the shelf life of fruits and vegetables, it is a well accepted method in developed countries like USA. The ripening process in bananas can be effectively delayed by irradiation at lower dose (0.2 kGy with a dose rate of 7.35 kGy). This effect is brought out through retardation of softening and colour change. Irradiation decreases sensitivity of the banana to its own endogenous ethylene without causing any phytotoxicity. At the same time, it does not affect ripening using high concentrations of exogenous ethylene (Mohapatra *et al.*, 2010).

Irradiation of bananas at 0.85KGy increased the storage life in the ripening room at 13°C from 14 to 29 days. Gamma irradiation of green bananas at 20-30 Krad delayed soluble solids formation, starch disappearance and respiratory activity. Although a number of countries in Asia irradiate fruits, the consumer acceptance of such a treatment is highly questionable. In some European countries like Germany, the irradiated fruits are not allowed.

12.13 Chilling Injury (CI)

Low temperature storage cannot be utilized to its full potential, in extending storage life of tropical fruits like mango and banana, because it is susceptible to chilling injury (CI) when stored at <13°C. Physiological events related to CI have been extensively described (Sevillano *et al.*, 2009). Modifications in membranes composition, especially the saturation degree of their constitutive lipids which regulate membrane fluidity and permeability, are the first notable effects of CI. An increased level of oxidative stress due to the overproduction of reactive oxygen species (ROS) has been considered a secondary effect of CI. The most common visual symptoms of CI in fruits are dark coloration of the skin, uneven ripening, development of off flavor and poor fruit quality. The symptoms of CI are strongly dependent on the duration and temperature of chilling treatment. Sometimes, symptoms may only develop when the produce is placed at higher temperatures and may appear immediately or the appearance may take several days. Chilling injury in bananas occur due to browning of cortical tissues at temperature below critical level (12.8°C). Several theories have been put forth with regard to the mechanism of development of chilling injury. Increase in the level of oxaloacetic acid in the fruits kept at lower temperature is one such earliest theories , but it is not clear whether this change is the result or cause of injury. Changes in membrane permeability due to exposure of tissues to chilling temperatures and electrolyte leakage have been reported to be responsible for manifestation of chilling injury in Cavendish banana. Variety, maturity, growing conditions and genomic constitution have been found to influence the chilling injury vulnerability of fruits (Narayana and Singh., 1998; Anon, 2009b). The symptoms of chilling injury appear to vary among banana cultivars and are related to the genomic group. The cooking banana cv. Bluggoe are most susceptible to chilling injury compared to the Pisang Awak (ABB). Narayana *et al.* (2002b) found that pome group of bananas were more susceptible than Cavendish or silk group. In general it has been observed that the genome of B group (B- *Musa balbisiana*) cultivars confers greater resistance to low temperatures compared with the AAA group (A-Musa acuminata). However, the Prata cultivar (AAB group) seems to be more tolerant to cold storage (around 12°C) than the Nanicão cultivar (AAA) (similar to Grand Naine), which makes it feasible to transport over long distances without impairing the quality of the fruit.

Lesser mature fingers are more susceptible to chilling injury than over mature fingers of banana. Higher sugar content was reported to confer resistance against chilling injury (Lyons,1973). Chilling injury in banana increases as the duration of exposure at 0°C increase from 5 to 20 days. Post harvest treatment with dimethyl polysiloxane, vegetable oil, mineral oil, polyamines, higher relative humidity, modified atmosphere storage, cold adaptation or cold conditioning, intermittent warming are some of the methods found to ameliorate the effect of chilling injury in banana (Wang *et al.*, 2006). Decrease in the activities of beta-amylase and increased polyphenol oxidase activity was found associated with chilling injury in banana (Nandini *et al.*, 2010). It is reported that the chilling injury can be reduced by cooling the fruit slowly, reducing the level of oxygen, and maintaining high relative humidity. However, the most practical method is to avoid exposure of banana to chilling temperatures below critical level which is determined depending on the sensitivity of the cultivar.

Chapter – 13

Ripening and Retailing

The series of modifications that transform a mature green fruit into a ripe fruit occur during a limited period of time and involve many different metabolic pathways. This means that they have to be strictly regulated and highly co-ordinated in order to lead to a successful result (Trainotti *et al.*, 2007). Fruits have classically been categorized based upon their abilities to undergo a program of enhanced ethylene production and an associated increase in respiration rate at the onset of ripening. Fruits that undergo this transition are referred to as climacteric and include tomato, apple, peach, and banana, whereas fruits that do not produce elevated levels of ethylene are known as non-climacteric and include citrus, grape, and strawberry. Climacteric fruits in general and bananas in particular are harvested while still green and firm but in un-ripened condition and then subsequently allowed to ripen. These fruits ripen naturally, but this may happen slowly, at a rate which is not predictable, and /or unevenly (Barry and Giovannoni, 2007).

Ethylene, the simplest unsaturated hydrocarbon, regulates many diverse metabolic and developmental processes in plants.Two systems of ethylene production have been defined in plants. System-1 functions during normal growth and development and during stress responses, whereas system-2 operates during floral senescence and fruit ripening. System- 1 is auto-inhibitory, such that exogenous ethylene application inhibits its synthesis. In contrast, system-

2 is stimulated by ethylene and is therefore auto-catalytic, and inhibitors of ethylene action inhibit ethylene production too. The biochemical features of the ethylene biosynthesis pathway in higher plants are well defined and have been reviewed by Bleecker and Kende (2000). Ethylene synthesized from methionine can be briefly explained in in three steps: (1) conversion of methionine to S-adenosyl-L-methionine (SAM) catalyzed by the enzyme SAM synthetase, (2) formation of 1- amino cyclopropane-1-carboxylic acid (ACC) from SAM via ACC synthase (ACS) activity, and (3) the conversion of ACC to ethylene, which is catalyzed by ACC oxidase (ACO). The formation of ACC also leads to the production of 5'-methylthioadenosine (MTA), which is recycled via the methionine cycle to yield a new molecule of methionine. Increased respiration provides the ATP required for the methionine cycle and can lead to high rates of ethylene production without high levels of intracellular methionine.

The time of onset of ripening in the absence of exogenous ethylene may be determined by a change in sensitivity to endogenous ethylene during maturation. In many fruits heterogeneity in ripeness and quality within commercial consignment is frequently observed as a result of variation in physiological maturities at harvest. Equalization of this is enabled both by acetylene and ethylene treatments in post harvest phase through initiation of simultaneous ripening. Ethylene releasing compounds are applied after harvest to initiate ripening of fruits. In the post harvest phase, endogenous ethylene production accelerates ripening and subsequent senescence of fruit tissues by stimulating physiological processes such as respiration, softening and chlorophyll degradation. Although ethylene is generally considered as the main triggering factor, an interaction between different hormones is necessary for the control and regulation of the fruit ripening process. The harvested bananas pass through three physiological development stages, namely the pre-climacteric or 'green life' stage, the climacteric or ripening stage which covers the eat-ripe, and finally the senescence stage when the fruits are over-ripe and dying.

13.1 Biochemical Changes During Banana Ripening

Ethylene, like in case of other climacteric fruits, plays a major role in the ripening of banana fruit. Normal ripening of banana fruit progresses from inside outward with pulp ripening proceeding the yellowing of peel. Colour changes during the ripening are due to alteration in chlorophyll and carotenoid contents of plastids or the accumulation of anthocynin in vacuoles. However, pigment changes in banana are confined to superficial cell layers of peel. The degreening of peel occurs as a result of chlorophyll breakdown by chlorophyllase (Thomas and Janave, 1992).The changes occurring in the banana fruit during ripening can be divided into those occurring in peel and those in the pulp. Most of the ripening related processes e.g. respiratory increase and autocatalytic increase in ethylene biosynthesis begin in pulp and persistent ethylene production by pulp is the major factor controlling peel yellowing. Similar to other fruits, many changes take place at the biochemical level during the course of ripening in banana, leading to the formation of the fully ripe fruit with characteristic aroma, taste and edibility.

13.1.1 Respiration Rate

Although the specific role of climacteric respiration in fruit ripening remains unclear, the recruitment of ethylene as a coordinator of ripening in climacteric species likely serves to facilitate rapid and coordinated ripening (Giovannoni, 2004). A rapid rise in rate of respiration concurrent with enhaned ethylene production is commonly seen in many climacteric fruits including banana (Song *et al.*, 1997; Sastry *et al.*, 2002). In banana fruit, respiratory climacteric stage is characterized by a massive conversion of starch to sugars in the pulp due to decrease in the activities of starch biosynthesis enzymes and increase in the activities of the enzymes involved in the starch breakdown and mobilization (Cordenunsi and Lajolo, 1995). A biphasic respiratory behaviour has been reported during ripening of banana fruit (Pathak *et al.*, 2003). The involvement of alternative respiration in thermogenesis (heat generation) during the ripening of banana (*Musa paradisiaca* var. *Mysore Kadali*) fruits, attached to a bunch, has been examined by Kumar and Sinha (1992) and it is suggested that alternative respiration may contribute to the temperature rise observed in ripening banana fruit. Low O_2 levels delay for long periods the onset of ripening in bananas (Hesselman and Freebarin, 1986; Mapson and Robinson, 1966; Quazi an Freebairn, 1970), and in initiated fruits slow down the rate of chlorophyll destruction and sugar accumulation. Even 10% O_2 delays the appearance of the yellow pigments (Brown, 1981).

13.1.2 Pulp to Peel Ratio

Various data are found in the literature indicating an increase in moisture content of the pulp of bananas during ripening. Of the physiological processes occurring, respiration causes the production of water. At the same time water is utilized in the hydrolysis of starch to sugar. As these two processes occur simultaneously, a part of the water produced by respiration may be used up in the hydrolysis of starch. It has been observed that the water formed by respiration and that utilized in the hydrolysis of starch, does not equal that used in hydrolysis. He also points out the probable changes in osmotic pressure with consequent transfer of water from peel to pulp. A report of United Fruit Company mentioned that osmotic pressure determinations of different parts of the bunch during ripening of bananas indicate that changes of pressure are such as to bring about a transfer of water from peel to pulp, as well as from the stalk through the crown and neck to the peel and pulp.

Banana fruit pulp increases in weight during the course of ripening as water is transported from peel and also from stalk and as a result of which the peel losses its weight. It changes the pulp to peel ratio termed as "Coefficient of ripeness" (Burdon *et al.*, 1993). The pulp to peel fresh weight ratio differs between cultivers from 1.18 to 2.28. Water loss, both to the atmosphere and to the pulp from the peel affects the post-harvest life of fruit at ambient tropical temperature. The pulp to peel ratio increases slowly in fruits held in cold conditions but is somewhat accelerated when fruits are placed in a ripening room. The increase in pulp/peel ratio indicates the stage of development of fruit. Pulp/peel ratio of 0.25 for immature banana, 0.37 for medium, 1.31 for mature and 3.1 for ripe fruit has been reported (Singh *et al.*, 1980).

13.1.3 Cell wall degradation

The banana fruit softens rapidly once ripening is initiated. Fruit softening is generally attributed to cell wall disassembly, particularly due to pectin solubilisation. Involvement of polygalacturonase (PG) or/and pectin methyl esterase (PME) in enzymatic disassembly of cellular wall has been widely reported. In addition, pectate lyase (PL), cellulose, b- galactosidase, transglycosylases or expansin proteins (Wang, *et al.,* 2006) may play a role in cell wall disassembly during fruit softening. The presence of carbohydrate rich cell wall is a distinguishing feature of a plant cell. The cell wall is approximately 30 % cellulose, 30 % hemicellulose, 35 % pectin and 5 % protein in dicotyledonous plants. In fruit cell wall, pectin content is higher and protein content lower. Pectins are the common and major components of primary cell wall and middle lamella, contributing to the texture and quality of fruits. During ripening, the insoluble 'pectins' decrease from 0.5% to 0.2% fresh weight with a corresponding rise in soluble pectins (Palmer, 1971). Cell wall disassembly in ripening fruit is highly complex, involving the dismantling of multiple polysaccharide networks by diverse families of wall modifying proteins (Nishiyama *et al.,* 2007). A class of proteins called expansins are cell wall-localized proteins associated with numerous changes in size and shape of cell during the developmental process (Cosgrove, 2000). Expansins appear to operate by disrupting hydrogen bonds between cellulose microfibrils and xyloglucans that bind them to one another in plant cell walls (Whitney *et al.*, 2000). Expansins have been shown to play an important role in fruit softening (Rose *et al.*, 1997; Anjanasree and Bansal, 2003).

Softening is associated with continuing progressive depolymerization of matrix glycan (Brummel *et al.*, 2004). Softening of the fruit is brought about by alterations in cell wall metabolism, which is due to shortening of polymer chain length, demethylation of carboxyl groups and deacetylation of hydroxyl groups in pectic substances. In banana, fruit ripening is also accompanied by solubilisation of cell wall including dissolution of the middle lamella of mesocarp which leads to softening of the fruit. These textural changes during ripening help in determining the shelf life of the fruit. Although in many fruits these changes occur gradually but in fruits like banana and strawberry they take place very rapidly and are the major cause of short post harvest life of these fruits. Fruit softening appears to be mainly associated with changes in the pectin fraction of the cell wall (Huber, 1983). The textural changes are accompanied by the cleavage of the methylester backbone (Carpita and Gibeaut, 1993), increased pectin solubilisation and reduction in pectin chain length (Seymour *et al.*, 1990; Smith *et al.*, 1990). The neutral sugar residues are also lost considerably from the pectin fraction of the cell wall (Wallner and Bloom, 1977). The ultrastructural and chemical changes are correlated with the *de novo* synthesis of various cell wall degrading enzymes (Fischer and Bennett, 1991; Dominquez *et al.*, 1992). PME plays an important role in determining the extent to which pectin is accessible to degradation by PG and thereby helps in cell wall expansion. Six forms of PME have been reported in banana (Markovic *et al.*, 1975) and total PME activity in pulp of 'Gross Michel' banana and 'Australian Cavendish' banana was found to remain constant during the ripening (De Swardt and Maxie, 1967). PG degrades demethylated pectins more actively than the methylated ones. In ethylene treated banana fruits PG activity increases as the fruit softens during the ripening (Agarvante *et al.*, 1990). Three multiple forms of PG have been isolated from ripe banana pulp, while unripe banana fruit contains only two multiple forms (Pathak and Sanwal, 1998). PG2 and PG3 increases during the ripening of banana fruit. It has been suggested that PG3, the endo-PG, catalyzes the conversion of de-esterified pectin molecules to the soluble smaller fragments of pectin, which subsequently is hydrolysed to D-galacturonic acid by PG2, the exo-PG. Increase in monogalacturonic acid in fruit pulp during ripening of banana fruit has also been reported (Asif and Nath, 2005). The activity of polygalacturonase during ripening in climacteric fruits has been positively correlated with softening of the fruit tissue and differential expression of its gene is suspected to be regulated by the plant hormone ethylene. Asif and Nath (Wade *et al.*, 1993) cloned four partial cDNAs, *MAPG1*, *MAPG2*, *MAPG3*, *MAPG4*. *MAPG3* and *MAPG4* are believed to be ripening related and regulated by ethylene, whereas *MAPG2* is associated more with senescence. *MAPG1* shows constitutive expression and is not significantly expressed in fruit tissue.

The role of PEL in fruit ripening has been demonstrated through molecular biology studies.Genes encoding PEL have been reported from banana. A clone, *Ban 17*, isolated from banana shows significant protein-sequence similarity to the pectate lyase of style and pollen of various plant species (Dominguez *et al.*, 1992). *Ban 17* mRNA reaches its maximum at the respiratory climacteric peak when changes in texture begin to take place. This pattern of mRNA accumulation suggests a role for the *Ban 17* protein in cell wall degradation during fruit softening. Two cDNA designated as MWPL 1 and MWPL 2 have been isolated from ripening banana fruit which are homologous to plant PELs (Pua *et al.*, 2001). During ripening, although transcripts sof both members are not detected in unripe preclimacteric fruits, they began to accumulate as ripening progresses and the level remaines high thereafter in overripe fruits. However, the magnitude of transcript accumulation differed between the two *pel* members, with substantially more abundant MWPL2 than MWPL1 in ripening fruit. Expression of both *pel* members is also affected by exogenous ethylene, whose presence stimulated accumulation of MWPL1 and MWPL2 transcripts in preclimacteric fruit, suggesting that ethylene may play an important role in regulating *pel* expression during banana fruit ripening. Recently, two distinct cDNA clones showing sequence homology to higher plant *pel* gene have been reported from banana fruit and both are strongly ripening related (Marin-Rodriguez *et al.*, 2003). It has been difficult to demonstrate PEL activity in plants, probably due to the presence of interfering substances in crude extract and low protein content. Employing a novel homogenizing medium, PEL activity has been recently demonstrated for the first time in banana fruit pulp (Payasi and Sanwal, 2003). PEL activity is not detected in preclimacteric fruit, but the activity increases progressively from early climacteric and reaches maximum level at climacteric peak and declines in post climacteric and overripened fruits. In banana, the climacteric peak corresponds with eating ripeness. The coincidence of PEL activity peak with the climacteric peak suggest the role of PEL in banana fruit ripening. The PEL activity in banana pulp tissue with a significant increase in calcium-dependent PEL activity during ripening has also been recently reported (Marin-Rodriguez *et al.*, 2003). The advancement of PEL activity peak in fruit treated with ethylene suggest the role of PEL in banana fruit ripening (Payasi *et al.*, 2004).

13.1.4 Starch, sugars, acids and phenolics

Starch constitutes the major carbohydrate in banana, which forms 20-25% of the fresh weight of the pulp of unripe fruit. During ripening this starch is degraded rapidly and the sugars like sucrose, glucose and fructose accumulate (Palmer, 1971). α-amylase, β-amylase, α-1,6-glucosidase and phosphorylase are the principal enzymes involved in starch degradation during ripening of banana (Seymour, 1993).

The conversion of starch to sugars is the most remarkable chemical change occurring in banana pulp during ripening (Palmer, 1971). During ripening the starch content declines from 20-23% in unripe fruit to 1-2% in fully ripe fruit and at the same time the soluble sugar increases from less than 1% to 20% (Agarvante *et al.*, 1990; Forsynth, 1980). Level of humidity during ripening was not found to influence the starch loss (Blankenship and Herdeman, 1995). Ripening of bananas is associated with an extensive breakdown of starch, the main carbohydrate reserve comprising 80 to 95% of the dry matter in the green fruit and 5 to 15% in the ripe fruit (Kanellis *et al.*, 1989), with a corresponding increase in the levels of soluble sugars, mainly sucrose, glucose and fructose. Accumulation of sucrose occurred earlier than glucose and fructose. Furthermore, the ratio of the increments of glucose over fructose was close to unity, indicating that the immediate precursor of these sugars is sucrose.

Amylose/ amylopectin ratio remains unchanged during the ripening (Garcia and Lajolo, 1988) but the concentration of maltose, sucrose, fructose and glucose increases during the ripening (Nabeesa and Unnikrishnan, 1988). Initially, sucrose is the predominant sugar and hexose sugar appear at the later stages of ripening and ultimately exceed sucrose concentration (Terra *et al.*, 1983). The peel contains much less sugar and starch than pulp but as the ripening proceeds the starch content of both tissues decreases while sugar content shows the increasing trend. Malic acid is the main acid component with substantial quantities of oxalic and citric acid in banana fruit pulp (Palmer, 1971).

During ripening the malic acid content increases, whereas the oxalic acid decreases. Total lipid content of banana fruit pulp as well as peel shows no significant change during the ripening (Goldstein and Wick, 1970). Banana is also a rich source of phenolics and tannins and ethylene treatment of banana fruits increases polyphenol oxidase activity and water soluble tannin contents. Tannins are translocated from peel to pulp during ripening, where they are metabolized by polyphenol oxidase. Polyphenols are largely responsible for the astringency of unripe banana and during the ripening there is a decline in the polyphenol content due to activity of polyphenol oxidase and peroxidase (Mendoza *et al.,* 1994). The characteristic aroma of banana fruit develops due to the production of wide spectrum of volatile compounds and specifically isoamyl acetate as a result of increased activity of alcohol transferase (Haard, 1967). The total protein contents of unripe banana pulp vary between 0.5% to 1.6 % and there is no significant change during the course of ripening (Steward, 1960). The polyphenol content of bananas harvested after 400 degree days remained unchanged during ripening, while bananas harvested after 600 and 900 degree days exhibited a significant polyphenol increase. Although dopamine was the polyphenol with the highest concentration in banana peels during the green developmental stage and ripening, its kinetics differed from the total polyphenol profile (Bonnet *et al.*, 2013).

Banana peel tissues are rich in phenolic compounds like 3,4-dihydroxyphenyl-ethylamine and 3,4- dihydroxyphenylalanine, which on oxidation by PPO forms brown pigments (Palmer, 1971). These phenols are also responsible for the formation of flavour volatiles. The banana pulp during ripening shows a marked increase in organic acids such as malic, citric and oxalic (Palmer, 1971; Marriott, 1980).

13.1.5 Aroma development

A wide variety of volatile compounds are formed during the ripening process which includes esters, alcohols, ketones, aldehydes and phenol esters. The result of principal aroma component analysis of banana in cold storage revealed that these compounds are more strongly affected during cold storage. Esters such as 2-pentanol acetate, 3-methyl-1-butanol acetate, 2-methylpropyl butanoate, 3-methylbutyl butanoate, 2-methylpropyl 3-methylbutanoate and butyl butanoate were drastically reduced in the cold stored Nanicão cultivar of banana (Facundo *et al.*, 2012).

13.2 Commercial Ripening

Fruit ripening involves a complex network of metabolic processes that include pigment, carbohydrate and aroma metabolism. The golden yellow colour of the ripe fruit is due to chlorophyll breakdown, which unmasks carotenoid pigments in the plastids. However, when ripening at warm temperatures above 24°C, bananas fail to develop a fully yellow peel as they retain high levels of chlorophylls in their peel (Blackbourn *et al.*, 1990). Thus, the peel of bananas remains a greenish yellow colour. These "green-ripe" fruits are perceived to be of poor quality and consequently fetch at a lower price. The response of banana fruit to temperature appears to be abnormal compared to the majority of other fruits. In plantain, chlorophyll breaks down in the peel is reported to be faster at higher temperatures within the range of 24–35°C. Chlorophyll breakdown in banana peel was exclusively suppressed at a temperature of 30°C. The inhibition of chlorophyll (Chl) breakdown under tropical temperature conditions in banana fruit has been studied in various aspects. Studies in the plastid ultrastructure showed that ripening at tropical temperature resulted in a retention of thylakoid membrane (Blackbourn *et al.*, 1990) which might prevent the release of chlorophyll from membranes for degradation (Hörtensteiner, 2006).

However, the biological basis for the effects of elevated temperature on thylakoid disassembly in banana is still unclear. Due to commercial consideration (long distance marketing) fruits are harvested at 75-80% which fail to ripen uniformly on its own and the market demands uniformity in colour, texture and flavour. To achieve this, exogenous application of ripening aids (calcium carbide, ethrel, ethephon and ethylene) are made use of by traders, both in domestic and export trade. Though many aspects of health safety/harzard of these chemicals are not clear, several experiments have been carried out in different parts of the world to study and standardize the dosage, time, temperature, method of application and effectiveness of these chemicals. Some of the significant findings with regard to its effects and interaction with other physiological parameters are summarized below.

Ripening in plantains is usually accompanied by complex biochemical and physiological changes in colour, flavour and textural parameters. Textural changes of bananas and plantains showed that when fruits were treated with ethylene 1000 ppm for 24 hours and then stored at different temperatures had reduced levels of textural profile parameters as compared to untreated control (Kajuna *et al.*, 1997). Ethylene spray or dipping at 500 ppm was found equally effective in inducing the ripening of banana cv. Champa. In respect of storage conditions, the results clearly showed that banana treated with ethylene when stored under conditions with adequate oxygen availability registered better colour development and required softness (Ghosh *et al.*, 1997). When Robusta bananas were dipped in calcium chloride solution (0-6%) and then stored at room temperature, the treatment advanced ripening by 2-5 days. After 14 days of storage fruits treated with 6% calcium chloride showed lowest firmness, highest titrable acidity, reducing sugars and total sugar content (Huddar *et al.*, 1990). Similar results were also reported by Narayana *et al.* (2002a).

Postharvest application of 1000-ppm ethephon enhanced ripening in directly treated fruits of banana cv. Williams but not in untreated fruits within the same bunch in storage (Shaabana, 1988). Fruits ripened at 20° or 30° C produced large quantities of volatile compounds including isoamyl acetate, but very little was produced at 35°C (Yoshioka *et al.*, 1982).

Mature green fruit of banana cv. Giant Cavendish stored at 40°C failed to ripen, apparently because of a reduction in protein synthesis. It also showed a reduction in the acid-phosphates activity at 40°C. Seven acid-phosphatase isozymes were shown to be present in mature green fruits and 3 more in fruits ripened at 20°C. But these were absent in fruits ripened at 40°C (Yashioka *et al.*, 1980). Postharvest dip of banana fruits in aqueous solution of abscisic acid and indole acetic acid significantly hastened ripening (as shown by increase in total sugars, acidity and ascorbic acid) during storage at 20°C. Treatment with GA and Kinetin however, retarded banana ripening as indicated by higher values for firmness, starch, cellulose and hemi-cellulose (Desai and Deshpande, 1975). Postharvest dipping of three-quarter ripe Dwarf Cavendish banana hands in 25 or 50-ppm 2,4,5-T retarded ripening at room temperature by 4-6 days. Treatment with 2,4-D at 25-125 ppm was less effective (Sadasivam and Muthuswamy, 1973). Unripe bananas were treated with 10-ppm ethylene gas for 1 hour and stored at 10 or 28°C (85-90% R.H.). Fruits at 28°C showed the characteristic climacteric peak after 6 days and ripened normally after 8-9 days. Fruits stored at 10°C had 42% lower respiration rate at 6 days and after 8-10 days it developed chilling injury characterized by brown discoloured and hard placenta. The chill injured fruits showed lower peroxidase activity and different patterns of peroxidase isozyme. It indicated that peroxidase present in ripening banana is affected by low temperature (Toraskar and Modi, 1984).

13.2.1 Use of ethylene gas or ethylene releasing compounds

Technically, ripening process can be induced by introduction of ethylene gas. Storage conditions such as relative humidity and temperature are very crucial in deciding the final quality of fresh fruit as they influence the physiological and chemical activity in the live biological commodities. It is reported that the relative humidity of above 80 % and temperature of about 20°C are the ideal for ripening of bananas.

In any given consignment, as there is wide heterogeneity in level of maturity of banana they need to be treated with a ripening aid like ethylene gas which is either applied through ethylene gas cylinders or catalytic ethylene generators or ethylene releasing compounds like CEPA (Ethrel, Ride, Ethephon, *etc.*).

Although dipping of fruits in diluted ethrel solution is recommended for enhancing ripening, it is a cumbersome process and may cause some problems if commercially available ethrel contains chemical impurities. Hence, the Food Safety and Standards Authority of India (FSSAI) has banned the application of ethrel solution in the form of dip or spray in 2009. However, use of ethylene gas at a concentration of 100 ppm is allowed. To overcome these problems ethylene gas is commercially used in modern ripening chambers. Catalytic reactors based ethylene generators are also available which produces ethylene gas using ethanol or methanol or ethrel. After 18-24 hours of exposure the fruits are taken out for completing the ripening process at room temperature or 18-24°C especially for Cavendish banana. The commercial ripening chambers follow a set procedure as described below for ripening of bananas.

Bananas are received at the ripening plant at color stage 1 (refer color chart). The product is pre-cooled to the pulp temperature of 18°C, after stabilization of temperature, ethylene is introduced at a concentration of 100-150 ppm. Most convenient and safe method of introducing ethylene is through ethylene generators. Cylinders containing 5% mixture of ethylene gas with nitrogen may also be used. After 24 hours, the chambers are vented. During this period the fruit starts producing ethylene by itself at the rate of 2-4 μl/kg/hr. Excess ethylene and carbon dioxide must be removed for faster and uniform ripening. For this purpose the chambers are vented after 24 hours of loading. Sometimes the ripening chamber suppliers provide the venting system along, which is capable of removing excess carbon dioxide in order to maintain concentration level below 10,000 ppm. Bananas are ripened to color stage 3 or 4 depending on mode of retailing for dispatch to market. Depending on cultivar and market requirements, ripening cycle may be spread from 4 to 8 days.

13.2.1.1 Color chart guide for ripening of banana

Mainly colour changes in banana during ripening is based on the peel colour rather than the pulp colour and hence colour of banana peel has been used in the assessment of the stages of ripeness of banana. Commercial standard colour charts are available in which there are 7 stages of peel color which are translated to a numerical scale where Stage 1=all green, 2= green with trace of yellow, 3= more green than yellow , 4= more yellow than green, 5= yellow with trace of green, 6= full yellow, 7= full yellow with brown spots. This color chart is more applicable for Cavendish bananas than any other.

These stages also indicate the level of ripeness and is a useful guide for commercial ripening service providers. The fruits are generally taken out of the ripening chamber at stage 3-4 so that it can be transported to the retail stores, where it will complete the ripening and becomes edible ripe in one or two days. For short distance transportations, which may involve 5-6 hours of journey by road, the fruits can be taken out at stage 2-3. For short distance transportations, which may involve 5-6 hours of journey by road, the fruits can be taken out at stage 2-3.

Recently non-invasive method of prediction of ripeness of banana was also developed based on capacitive property of banana fruit. Relative permittivity was correlated with quality parameters of banana fruit. Green-ripe banana fruits have larger permittivity than the full-ripe ones, the permittivity of which was decreased as a result of the ripening stage. A quadratic regression equation had higher prediction power to assess the ripening of banana (Soltani *et al.,* 2010; Tapre and Jain, 2012).

Shaaban (1988) reported that postharvest application of 1000-ppm ethephon enhanced ripening in directly treated fruits of banana cv. Williams but not in untreated fruits within the same bunch in storage. Hussein *et al.* (1985) reported that 'Hindi' and 'Paradica' varieties of bananas coated with 1.5% Prolong (sucrose ester) delayed the ripening significantly. Banks (1985) reported that coating Giant Cavendish banana fruits with Pro-long 24 hours after initiation of ripening decreased their skin permeability to gases and depressed their oxygen content. It also suppressed rapid ethylene production and climacteric rise. Coating delayed the chlorophyll loss, which normally accompanies ripening.

13.2.2 Other Methods of banana ripening

Other conventional methods of ripening of bananas like smoking, burning of kerosene stove, use of incense sticks, etc. in closed chambers are also found to release ethylene or acetylene or mixture of both which aid in uniform ripening of fruits. However, in these methods other harmful gases like carbon monoxide are also released. The use of ethylene gas has been proved to be more beneficial over acetylene gas for the purpose of ripening of fruits by several research workers. However, both acetylene and ethylene being explosive gases, mishandling or inappropriate usage in large quantities can cause harm or accident.

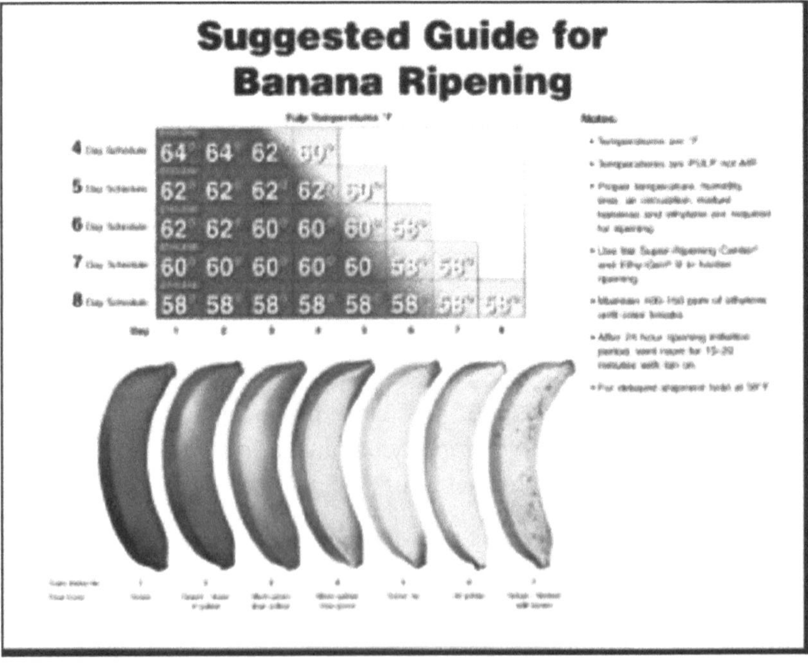

Flow sheet of postharvest operations for Export of Banana

Selection of bunches
↓
Cutting of bunches
↓
Holding of bunches and putting on stretches
↓
Carrying of bunches to packhouse
↓
Dehanding of bunches
↓
Deflowering of banana hands
↓
Giving proper cut to crown end or trimming
↓
Giving hands fungicidal solution dip
↓
Drying the surface moisture of hands
↓
Placing hands in boxes
↓
Putting ethylene absorbent sachets in the boxes
↓
Weighing of boxes and adjusting the weight
↓
Transport boxes to pre-cooling units
↓
Pre-cooling and cold storage
↓
Transporting in refer containers
↓
Shipping

Chapter – 14

Minimal Processing

By definition, minimal processing would encompass any procedure, short of traditional complete preservation procedures (heat sterilization etc.) that adds value to the fruit or vegetables (Floros, 1993). Other terms used to refer minimally processed products are lightly processed, partially processed, fresh processed, fresh cut and prepared. Minimally processed fruits and vegetables are products those have the attributes of 'fresh-like' quality. A number of minimally processed products of fruits and vegetables are now being produced like peeled and sliced potatoes, shredded lettuce and cabbage, washed and trimmed spinach, chilled peach, mango, banana, melon and other fruit slices, vegetable snacks such as carrot and celery stick and cauliflower and broccoli florets, packaged mixed salads, cleaned and diced onion, peeled and cored pineapple, fresh sauces, peeled citrus fruits and microwaveable fresh vegetable trays. Minimal processing of fruits and vegetables has two purposes. First, it is important to keep the product fresh but convenient without losing its nutritional quality. Second, the product should have a shelf life sufficient enough to facilitate distribution within the region of consumption. The microbiological, sensory and nutritional shelf life of minimally processed fruits and vegetables should be at least 4-7 days but preferably even longer (Ahvenainen, 2000).

The minimally processing operations of fresh-cut fruit and vegetables mainly consist of simple unit operations including cooling, washing, trimming or shredding, peeling, grating, slicing, drying *etc*., with mild preservation techniques before packaging which make the product ready-to-use. The cells of minimally processed fruits and vegetables are often physiologically active, resulting in on-going protective and/or deteriorative changes in cell structure and cell chemistry, as well as alteration in the chemical composition of the environment (Huxsoll, 1989). Minimal processing increases the perishability rather than the stability of the fruits and vegetables (Rolle and Chim, 1987; Shewfelt, 1987). MP products possess higher rate of respiration, which generally leads the ageing of the products by using the energy reserve during oxidative-reduction process (Ryle, 1984; Watada *et al.*, 1986). However, the rate of respiration remained unaffected after peeling and slicing in ripe banana (Watada, *et al.*, 1990).

Other types of deterioration which occur include chemical and enzymatic changes and microbial deterioration (Lee *et al.*, 1995; Watada *et al.*, 1996). The shelf life of minimally processed fruits and vegetables is generally limited by changes in their sensory properties and not by microbial growth (Jacxsens *et al.*, 2002; Jacxsens *et al.*, 1999; King *et al.*, 1991). Minimally processed fruits and vegetables are preserved by refrigeration (Feinberg *et al.*, 1987; Huxsoll and Bolin, 1989; Shewfelt, 1987), chemical preservatives, additives, bio-preservatives, mild heat treatments, microwave processing, reduction of water activity, ionizing irradiation and, disinfectants (electrolyzed water treatment, chlorination, hydrogen per oxide). Edible coatings (multilayer coating, osmotic membrane coating) are also being used in minimal processing of fruits and vegetables. Production of fresh-cut fruits is becoming an important task in food industry because of their convenience as ready-to-eat products as well as for the health benefits associated with their consumption.

Fresh-cut banana slices have a faster rate of browning and softening because of wounding (Vilas-Boas & Kader, 2006). Degree of browning depends on phenol content, poly-phenol oxidase (PPO) activity (Zawistowshy *et al.*, 1991) and oxygen concentrations. Over the years, different physical and chemical techniques have been developed to extend the shelf life of fresh-cut produce: refrigeration, disinfection (Hong & Gross, 1998), ethylene absorbers (Abe & Watada, 1991), gamma irradiation (Chervin & Boisseau, 1994), edible coating (Baldwin *et al.*, 1995), chemical dipping (Vilas-Boas & Kader, 2006) and controlled/modified atmosphere (Brecht, 1999). Because different fruits react differently to different treatments, it is imperative that the right combination of these treatments is determined in order to extend its shelf life.

During the last decade, minimally processed high moisture banana fruit products (HMFP), which are ambient stable (with $a_w > 0.93$), have been developed in seven Latin American countries, under the leadership of Argentina, Mexico, and Venezuela.

Application of calcium compounds helps to maintain firmness of fresh-cut banana (Vilas-Boas & Kader, 2006). A 2-min dip in a mixture of 1% (w/v) calcium chloride + 1% (w/v) ascorbic acid + 0.5% (w/v) cysteine effectively prevented browning and softening of banana slices for 6 days at 5°C. Dips in less than 0.5% cysteine promoted pinking of fresh-cut banana slices, while concentrations between 0.5% and 1.0% delayed browning and extended the post-cutting life to 7 days at 5°C (Vilas-Boas & Kader, 2006). Furthermore, fresh-cut banana treated with chemical dip at pH 2.5 showed a longer shelf life in terms of firmness than pH 7.0. MA helps to maintain freshness of fresh-cut produce by inhibiting metabolic activity, ethylene sensitivity and production, and/or physiological and pathological deterioration during storage (Gorris & Tauscher, 1999). However, level of oxygen and/or carbon dioxide should be regulated as respiration of the product becomes anaerobic when oxygen levels decline too much (McHugh & Senesi, 2002), leading to accumulation of ethyl alcohol, or anaerobic metabolism (normally below 1% O_2) that leads to off – flavours (Purvis, 1983). Combination of 2–5% O_2 and 2–5% CO_2 was found to delay ripening and reduce respiration and ethylene production rates of whole banana fruit (Kader, 2005).

Functional, nutritional, organoleptic and mechanical properties of coatings can be improved by the use of additives like antibrowning agents, preservatives, firming agents, plasticizers, nutraceuticals, volatile precursors, flavours and colours (Baldwin *et al.*, 1996; Vojdani and Torres 1990). Glycerol and polyethylene glycol are the most often used plasticizers for cut fruits (Wong *et al.,* 1994). These substances have the ability to modify the mechanical properties of the coatings by moving polymer chains apart and reducing the rigidity of the structures (Guilbert & Biquet, 1996). Some studies have been conducted on the effect of combination of chemical dip and/or CA and/or coating. Vilas-Boas and Kader (2006) studied the effect of combination of chemical dip and atmosphere composition on fresh-cut banana. They found that combination of CA (2 kPa O_2 + 10 kPa CO_2) with chemical dip (1% $CaCl_2$, 1.0% ascorbic acid and 0.5% cysteine) did not prevent changes in firmness of fresh-cut bananas further than the chemical dip alone, while Aguayo *et al.* (2008) also found that combination of calcium dip (1% w/v $CaCl_2$) + CA (3 kPa O_2 + 10 kPa CO_2) had no additional effect on firmness retention of fresh-cut strawberries compared to these two treatments when provided separately. Combination of edible coating, chemical dip and/or CA may then be useful in extending the shelf life of fresh-cut fruit, banana, in particular. There is little information about the combined effect of edible coating and/or chemical dip and/or CA on fresh-cut fruits. Moreover, only few research works have been conducted on freshcut banana.

Vilas-Boas and Kader (2001) reported on the effect of 1-MCP treatment on fresh-cut bananas, kiwifruits, mangoes, and persimmons. Banana did not respond when whole fruit were treated before cutting (TP), but a postcutting application reduced softening of slices (Vilas- Boas and Kader, 2001, 2006). 1-MCP applications could be useful in situations in which raw product is handled alongside ethylene generating produce (e.g., tomatoes, bananas, or apples) before processing or if the cut leafy products are packaged with ethylene-generating products as part of a specialty mix. One area that has only been explored more recently is the use of 1-MCP as a co-treatment with treatments that control aspects of quality change other than ripening. There are a few papers demonstrating that 1-MCP, when combined with other postcutting treatments, has a synergistic effect on quality retention and control of decay (Aguayo *et al*., 2006; Luo, 2005; Mao *et al.,* 2006; Toivonen and Lu, 2006).

The combined effect of chemical dip and/or edible coating and/or controlled atmosphere (CA) on quality of fresh-cut banana was investigated by Bico *et al.* (2005). Banana slices were subject to a 3-min dip into a solution containing 1% (w/v) calcium chloride, 0.75% (w/v) ascorbic acid and 0.75% (w/v) cysteine and/or combined with a carrageenan coating and/or combined with controlled atmosphere (3% O_2 + 10% CO_2). Physico-chemical and microbiological qualities were evaluated during 5 days of storage at 5°C. Dip combined with CA treatment prevented product weight loss and increase of polyphenol oxidase activity during the 5 days of storage. Colour, firmness, pH, tritatable acidity and total soluble solids values and total phenolic content presented the smallest changes. Microbial analysis showed that minimally processed bananas were within the acceptable limits during 5 days of storage at 5°C.

Chapter – 15

Banana Processing and Processed Products

Fresh bananas are widely consumed in developed countries and the use of processed banana products is not as high. This is because of the fact that fresh bananas are readily available round the year (Sole, 1996). Though there are several processed banana products that are being consumed and traded all over the world, except for the puree and ripe fruit powder, other products are mostly in small or medium scale sector and mostly limited to banana producing countries. Banana figs produced by sun drying of ripe bananas are one of the oldest processed products besides other popular processed products like dehydrated flakes and flour.

In banana growing states in India, though, banana is produced in substantial quantities, losses are found to occur at different stages of handling after harvest. During the market glut, the prices crash down and the farmers suffer heavy losses due to the distress sale. Due to poor transportation and storage facilities, a sizeable quantity of this fruit is wasted due to its perishable nature. Great potentialities exist for reducing the post-harvest losses by value addition through processing of banana. In a study conducted to assess the extent of value addition in banana in Jalgaon district of Maharshtra, it was found that the usual value added products produced are wafers, chivda, powder, etc. It was also found that value addition in banana is augmenting the farmers' farm incomes, and helping in sustainable development process of their rural economy (Bhalerao *et al.*, 2010).

Bananas containing higher starch (green or unripe stage, and particularly cooking bananas) and plantains are preferred for processing than ripe bananas. Cooking bananas can also be used for beer, wine, flour, and in crisps/chips particularly in African and Latin American countries (Robinson 1996; Akubor *et al.*, 2003; Aurore *et al.*, 2009). Plantain is less prone to browning than banana hence it retains a stable orange color on being cooked and does not easily undergo maillard reaction prior to cooking (Ngalani *et al.*, 1993; Yang *et al.*, 2004).The dessert banana can serve as raw material for industrial processing and its potentials includes the production of fruit juice, purees, starch, fermented and fruit drinks, marmalade, jam, ice creams, pastry ingredients, confectioneries, and sorbets.

In Philippines 18% bananas goes to the processing sector, and 49% goes to the local market for making various food items (http://platforms.inibap.org/ processing/ images/stories/file/pdf/philippines.pdf accessed on 20.11.2013) . The Philippines and Thailand in Asia are two countries which export large quantities of processed bananas. The Philippines chip exports amounted to 28,000 tons worth about US$40 million in 2010. The top destinations were Vietnam, which bought 20 percent of the total export volume during the year, followed by the US, China, Germany, and the United Kingdom. In the Philippines Saba variety is widely used for producing products like chips and Catsup. Its exports actually reduced between 2006 to 2010 by seven percent per year, largely due to problems on the supply of Saba variety of banana.

Brazil, Honduras and Mexico are the main countries exporting dried or desiccated or evaporated bananas to United States. Banana puree or mashed bananas are at present the highest volume processed banana product in the world. Banana pulp, banana flour, powder, chips, canned banana slices, banana jam, banana wine and juice are the other products popular in Taiwan (Tsen *et al.*, 2003)

The biggest export market for processed banana products in the world is USA, followed Canada, Saudi Arabia and United Arab Emirates. The banana flour, meal and powder from the Philippines were exported solely to Japanese market at a volume of 2.2 tones valued at US$14, 771 in 1998 (Narayana and Pillay, 2011).

Internal consumption and availability of bananas for processing are hard to estimate particularly in countries such as India. Though the internal demand is tremendous for banana in India, due to poor infrastructure and faulty postharvest handling practices the amount of postharvest losses to an extent of more than 5.0 million tonnes. The fruits and vegetable processing industry in India is steadily increasing and presently around 2.2% of its total production is processed. However, this sector has been showing a constant increase in its export performance. With constant increase in the productivity there is going to be huge volumes of marketable surpluses, which provide enormous scope for development of banana processing industry in India. The major products processed from banana in different countries including India are enumerated and discussed below.

15.1 Banana Chips/Crisps

Banana chips manufacture is a widespread activity in several banana growing countries across the world. In India, most popular processed banana product is Nendran (French Plantain) chips, which has grown from a household art to a well established cottage industry in states of Kerala and Tamil Nadu. Banana chips or crisps are manufactured form 'Nendran' variety in India while 'Saba' variety is used mostly in Philippines. It can be made by frying in oil or dried in sun for later usage as flour. Suvittawat and Babpraserth (1996) studied banana chip production using nine banana cultivars with different genomes. The consumers preferred chips from 'Khai' cultivar the most. Salted plantain chips are made by deep-frying raw banana slices in suitable cooking medium and salting them. The Bluegoe type cooking banans are also used in India for making chips. Coconut oil is the most preferred medium in Kerala and Tamil Nadu, while in other countries cotton seed oil or corn oil is used. Sugar coated banana chips is another form of chips which has good demand in several parts of the world. Stage of maturity and use of antioxidants play a very important role in its quality and storability. Several studies have been done to identify the suitable variety of banana for chips making (Narayana *et al.,* 2002d; Adeniji and Tenkouano, 2007; Molla *et al..*, 2009). While Nendran is the most preferred variety, Zanzibar, Mindoli, Robusta are the other alternatly suitable varieties. Chips are coated with cane sugar and refried to produce sugar coated banana chips. Chips are also coated with banana puree and sugar before frying the second time. Hydrocolloids such as alginate, CMC and pectin were shown to decrease the amount of oil uptake during frying. Excess oil can also be removed from chips by using basket centrifuge. Diaz *et al.*, (1996) observed that chip quality can be improved if frying temperature is maintained at 167°C.

Badia (1987) reported that degree of ripeness of plantain is important in making chips and that browning occurs if the sugar content is higher than 1% in the fruits. Therefore, the stage of ripeness is very important parameter to be considered while selecting the fruit. However, in some parts of India, semi-ripe fruits are also used to produced sweet and sour chips. Shere *et al.* (1993) studied the effect of maturity stage on quality of fried banana chips. 'Basrai' variety of banana was harvested either 85 or 100 days after flowering. Maturity stage of 100 days after flowering was not good for preparation of chips because of high sugar content, which caramelized at high temperature. Addition of antioxidants like propylene glycol, propyl gallate and citric acid have been recommended for reducing the rancidity (Badia, 1987; Ogazi, 1996; Narayana *et al.*, 2002d). Effects of systems for contact of banana slices (solid phase) and palm oil were studied during deep frying of banana chips by Totte *et al.* (1996). The results showed that shaking homogenizes the temperature of the oil and enhances mass and energy transfer. Exchange of matter and heat took place during the first three minutes of cooking. Water loss was less than 5g/100g of raw material and gain in lipids was less than 16g/100g. Optimum cooking time was observed to be 200 seconds with a minimum ratio of energy consumption to evaporation of water. Kachru *et al.* (1996) reported that a rotary electrical machine for cutting raw banana into chips has been designed and manufactured which has a slicing throughput of 61 kg per hour. Its efficiency is 50% higher than conventional chip cutting method.

Under room temperature (25-30°C) and super market temperature (20-25°C) good quality product could be stored up to 1-month and 3-months respectively. The shelf life of chips depends on packaging material and storage environment. Traditionally the LDPE or PVC film bags are used, however, the shelf life of product in these films are 3-4 weeks only due to high level of permeability to oxygen and moisture. In Nigeria cellophase has been recommended as the packaging material for plantain chips (Ogazi, 1996).

Many plantain varieties are rich in carotenoids. It has been reported that chips made from certain banana varieties could contribute recommended daily allowance of retinol, iron and zinc (Adeniji and Tenkouano, 2007). Bresnahan *et al.* (2012) studied the retinol bioefficacies of unripe and ripe bananas (study 1A), unripe high-provitamin A bananas (study 1B), and raw and cooked bananas (study 2) in retinol-depleted Mongolian gerbils (n = 97/study) using positive and negative controls. Their findings suggested that thermal processing greatly improved the retinol bioefficacy than ripening.

15.2 Banana Puree/Pulp

Banana puree is the most important product in international trade and is largely used for preparation of juice/ beverages, jam, baby food, etc. Earlier banana puree was prepared and packed in cans by aseptic canning technique, but after the advent of asceptic processing it is packed in aseptic laminate bags. The process of banana puree preparation involved peeling the ripe fruits, immersion in sodium bisulphate solution (1.5% for 3 minutes), draining, milling in a fruit mill having 1/8" sieve, pureeing or pulping, passing through a plate heat exchanger at 93.3°C for 1 minute, cooling to ambient temperature. Then it is passed through a finisher having 0.033" sieves to remove the small seeds and fibrous material. The pH of the finished pulp is adjusted to 4.1 using citric acid. Finally the preservative (potassium sorbate at 200 ppm) is added and filled into aseptic bags under steam and sealed. The bags are further put into plastic barrels before storage at low temperature (Sole, 1996; Occena-Po,2006). Pre-treatments of plantain by blanching in hot water, steam, microwave treatment or calcium chloride modified the texture and the organoleptic characters. Calcium chloride and microwave pre-treatments hardened frozen banana pulp. Microwave pre-treatment seemed to be the most suitable for frozen banana (Giami, 1991). Garcia *et al.* (1987) conducted a study to reduce the time required to lower the pH of banana puree by lactic acid fermentation. The best production conditions were found to be 37°C and 10% (W/W) of inoculum, the condition under which the pH decreased from 4.8 to 4.3 in approximately 8 hours.

15.3 Banana Flour

The production of banana flour is widespread in African countries (Uganda, Kenya, Tanzania) which is mainly used for various household product making (Narayana and Pillay, 2011). Banana flour is utilized in infant foods, beverages, in baking and confectionary industries. Production of banana flour is carried out by peeling and slicing green fruits, exposure to sulphur dioxide gas or dipping in sulphurous acid and then drying in a countercurrent tunnel dryer for 7-8 hours with an inlet temperature of 75°C and outlet temperature of 45°C, to a moisture content of 8% and finally milling. The full three-quarter-maturity stage of fruits contains right proportions of starch and sugar, which results in good quality flour. If under mature fruits are used the resultant flour tastes bitter and astringent due to high tannin content. This product has been made for export in Ecuador, Colombia, Canada and Switzerland generally in small volumes. Trobana in Ecuador also produce green banana flour for export market. Steaming of bananas prior to making flour is reported to increase water uptake, density and solubility of banana flour and resulting in pastes of low bulk density which is a desirable quality in weaning foods and food supplements (Muyonga *et al.*, 2000). Banana flour can also be made by freeze drying process (Mota *et al.,* 2000). The traditional way of making banana flour is to peel, cut into small pieces and sun drying for few days. The dried pulp is then ground into powder in a mortar and pestle. Flour produced in this manner is usually brown due to the effect of enzymes such as polyphenol oxidase snf prtoxifsdr ot oyhrt non-enzymatic reactions (Cano *et al.*, 1997). Steaming the bananas has been used to reduce the enzymatic browning (Suntharalingam and Ravindran, 1993).

The banana flour can be used as an adjuvant in several food preparations and baby food formulations. It can be blended with other cereal flours for making chapaties, biscuits, bread, papad and extruded foods.

Green banana is very rich in starch and its flour may contain (61.3–76.5) g/ 100 g of starch (dry weight basis.) and 6.3–15.5 g/100 g of fiber content (Juarez-Garcia, *et al.*, 2009; Mota *et al.*, 2000). The banana starch is reported to be the resistant starch type 2 (RS2) (Faisant *et al.,* 1995). Such starch is found only in raw potatoes and green bananas, whose crystalline structure makes it less susceptible to hydrolysis by amylase enzyme (Gonza′ lez-Soto *et al.*, 2007).

Resistant starch has attracted interest because of its positive effects in the human colon and implications for health (Langkilde *et al.*, 2002; Englyst *et al.*, 1992). During the process of ripening this starch disappears rapidly due to the action of several enzymes acting in tandem. In recent years, there has been a considerable interest in the possibility of improving control of diabetic patients by altering the glycemic impact of the carbohydrates ingested. A tool for ranking foods with respect to how they potentially raise blood glucose is the glycemic index (GI) concept (Jenkins *et al.,* 1981), which leads to the preferred selection of "slow carbohydrate" food items/diets (Jenkins *et al.,* 1981; Bjorck *et al.*, 1994). Though it cannot be regarded as a general rule, a nutritional variable that may be linked to low GI properties is the RS content of a food or complete meal (Truswell, 1992). Against this background potential of banana starch makes it an ideal candidate for developing various kinds of speciality / functional foods.

Aparicio-Saguila *et al.* (2007) prepared a resistant starch-rich powder (RSRP) from autoclave-treated lintnerized banana starch. The rate of starch digestion in vitro was studied using the cookies made from RSRP and compared with those prepared from wheat flour (control). The hydrolysis index (HI)-based glycemic index for the RSRP-cookies was 60.53, which was significantly lower than for control samples (77.62), suggesting a "slow carbohydrate" feature for the RSRP-based cookies. These results suggested that RSRP from banana starch as a potential ingredient for bakery products containing slowly digestible carbohydrates.

Martinez-Ovando *et al.* (2009) explored the possibility of use of unripe banana flour as a food ingredient to make pasta (spaghetti) of high quality, on the basis of low-carbohydrate digestibility, and increased resistant starch and antioxidant phenolics contents. Formulations consisting of 100% durum wheat semolina (control) and mixtures of semolina:banana flour of 85:15, 70:30 and 55:45 were prepared for spaghetti processing. Nutritional composition, in vitro kinetics of starch digestion and antioxidant capacity were determined. The addition of banana flour increased the indigestible fraction and the content of phenolic compounds in the spaghetti. As a consequence of the compositional changes, a slow, low rate for the enzymatic hydrolysis of carbohydrates was observed. Moreover, banana flour spaghetti possessed increased antioxidant capacity.

Thirty per cent substitution of wheat flour with Banana Flour (*Musa acuminata* X *Musa balbisiana* Colla cv. Awak) resulted in significantly higher total dietary fibre (TDF), and especially insoluble dietary fibre (IDF), resistant starch (RS) and total starch contents of noodles. Thirty percent of BF significantly improved the antioxidant properties (AP) of noodles in terms of the total phenolic (TP) content and inhibition of peroxidation (Choo and Aziz, 2010).

15.4 Banana Powder

Banana powder is made using fully ripe banana pulp. The pulp is made into a fine paste by passing through a colloidal mill. Blanching is an important step in banana powder processing for controlling discoloration in the product. About 1-2% potassium metabisulphite solution is added to improve the colour of final product. Further the pulp is spray or drum dried and the solids are recovered. In the latter method the moisture content is reduced to 8-12% and it is further decreased to 2% by drying in tunnel or cabinet dryer at 60°C. The dry powder is highly hygroscopic and is packed in bag-in-box packaging. The bags are made from laminated material with good moisture barrier layers. This product is of high market value as it is used in confectionary industry, ice creams, weaning foods and baby foods. Powder from ripe banana fruits has potential for several industrial and domestic uses (Abbas *et al.,* 2009). Ripe banana powder is more soluble and sweet compared to flour. It can be incorporated into food products which are required to be sweet. As it has a high sugar and low starch content it can be used as a substitute for fresh banana in making traditional cakes or their premixes as well as in the processing of banana snacks, crackers or crisps. The "Bhimkol' (ABB) variety of banana found in northeastern hill regions of India is rich in macro and micronutrients like K, Mn, Zn and Se. Traditionally it is used for baby food in that region (Borthakur and Arnold, 1990).

15.5 Banana Flakes

Banana flakes can be made by several methods both from ripe and semi-ripe banana fruits. In drum drying method banana puree is coated on to large chrome-plated drum dryers. The puree, extracted in similar way to that of banana puree is fed directly to the dryers. The water evaporates as the drum turns, forming a film of dried material on the drum surface. Carefully adjusted knives remove the dry film as a continuous sheet, which passes to an air-conditioned room where it is broken in to flakes. It is then sifted and packaged in bag-in-box containers (Sole, 1996). Trobana is the world's largest supplier of banana flakes. Freeze-drying and foam mat drying are other techniques for producing banana flakes.

The technique of foaming has proved effective in creating a porous structure, which is an important requirement for crisp food. The foaming of ripe bananas and the forced air-drying characteristics of the resulting banana foam mats is another method studied by Sankat *et al.* (2004). Fresh banana puree with a density of 0.93 g/ml was foamed to a density of 0.50 g/ml after 12 min of whipping by the addition of 10 g/100 g soya protein as a foam inducer. Of various foaming agents used glyceryl monostearate did not induce foaming while commercially available food ingredients, Dream Whip and Gelatine induced foaming but such foams were not suitable for subsequent drying. Banana foam mats were dried at temperatures from 45°C to 90°C in a forced air, cabinet dryer, to a hard, porous and brittle solid which was amenable to grinding so as to produce a dehydrated banana powder. The drying time (*t*) was directly related to the thickness of the foam mats. *K* values increased from 0.248 to 0.809 h^{-1} as the drying air temperature was raised from 45°C to 90°C. Increasing the air velocity from 0.62 to 1.03 m/s did not profoundly influence the drying rate.

Foam density and the type of foaming agent indeed play a key role in determining the drying kinetics and textural property of the foamed food. The influences of the foam density and the types of foaming agents on the moisture diffusivity as well as the quality in terms of microstructure, texture and volatile losses of banana foams were investigated by Thuwapanichayanan *et al.* (2012). When three foaming agents, i.e., fresh egg albumen (EA), soya protein isolate (SPI) and whey protein concentrate (WPC) were used the results showed that WPC banana foam could retain more open structure during drying. This morphology provided less shrinkage and led to higher values of the effective diffusivity as compared with that of SPI and EA banana foams. In terms of the textural properties, WPC and EA banana foams were spongy and less crisp than SPI banana foam. Samples with lower foam densities exhibited higher values of the effective diffusivity, smaller hardness and lower crispness than those with higher foam densities. However, disadvantage of this method is loss of volatile substances during foaming step.

The pre-treatments prior to drying can also influence the quality of banana flakes. The effects that four pretreatments (blanching, chilling, freezing, and combined blanching and freezing), used prior to drying, had on the drying rate and quality of bananas were investigated by Dandamrongra *et al.* (2003). An untreated sample was used as a control. The bananas were dried at 50 °C in a heat pump dehumidifier dryer, using an air velocity of 3.1 m s^{-1}, until a final moisture content of approximately 25% dry weight basis was attained. While the initial drying rate was highest for the blanched treatment, the two pretreatments involving freezing resulted in the shortest drying times. The blanched sample was most preferred in terms of colour while the frozen samples exhibited extensive browning. The texture and flavour was significantly reduced in all samples that involved blanching and/or freezing.

Wang *et al.* (2007) studied the volatiles of dehydrated banana powder produced by different drying methods. Banana puree was dehydrated by using three different drying methods: vacuum belt drying (VBD), freeze-drying (FD) and air-drying (AD) to produce banana powder. Volatiles were extracted from dried banana powder with solid phase micro-extraction (SPME), and separated and identified by gas chromatography–mass spectrometry (GC–MS). The main components detected in banana powder which are responsible for its fruity odor were 3-methylbutanoic acid 3-methylbutyl ester, 3-methylbutyl acetate and butanoic acid 3-methylbutyl ester. Most of the alcohols identified in banana powder were enols and some were of the long chain type. Eugenol and elemicin which give the product its typical mellow aromas were also identified. Alkyls, alkene and alkyne constituted the minor components in banana powder. Based on the principal component analysis (PCA) of statistical analysis system, it was inferred that the preferred method for producing banana powder with the optimum aroma is FD, followed by VBD and then AD. Both vacuum belt drying and freeze drying methods are complex processes requiring absolute control of vacuum and temperature.

Often due to several reasons, the investments in higher capacity machinery becomes prohibitive. Under such circumstance, small scale processing or domestic processing could be encouraged. Drying tests for ripe and green banana (*Musa* cv. Giant Cavendish AAA) slices and for a foam made from ripe banana puree were performed in a household microwave oven and in a laboratory forced draft warm air oven (Garcia *et al.*, 2007). Weight change of the banana slices as a function of time in both ovens, and temperature and relative humidity (RH) of air leaving the microwave oven were recorded. Drying rates were calculated and the data fitted to a variable diffusion model. Drying time in the microwave oven was seventeen to twenty times less than in the forced draft oven. The drying energy efficiency in the microwave oven was about 30%.

15.6 Banana Figs or Dehydrated Banana

Banana figs are dried or dehydrated banana with sticky fig like consistency and very sweet taste. The product is made from fully ripe fruits. Varieties with high total soluble solid content (19%) are highly suitable for making figs. Banana drying is of considerable interest for food supplies in Ethiopia. Batches of slices are treated for 5 minutes in boiling 50% sugar solution and then exposed to sunshine in black boxes with glass covers (Osama Ahmed Ismail and Asrat, 1985). Banana-passa is a product manufactured in Brazil by drying 'Nanica' bananas. Drying was studied at three temperatures (50, 60 and 70°C) and at three air velocities (0.5, 1.0 and 1.5m/second). The best conditions for obtaining a quality product were a temperature of 60 or 70°C and air flow of 1.5m/second (Nogueira and Park, 1992). In Uganda, 'Sukali Ndizi' (syn. 'Apple banana', AAB genome) are dessert bananas that are processed into dried slices for export. It is exported to Europe where the prices of dried banana slices are strongly depended on the quality and certification of the product. While this variety is being devastated due to Fusarium wilt, the alternate variety suggested is 'FHIA-01' the quality of which is atleast as good as 'Sukali Ndizi' (Van Asten *et al.*, 2010).

Osmotic dehydration has received greater attention in recent years as an effective method for fruits and vegetables preservation. Being a simple process, it facilitates processing of tropical fruits like banana, sapota, mango, pineapple etc, with retention of their initial fruit characteristics viz. colour, aroma and nutritional compounds. Osmotic dehydration is a process of water removal from fruits, because cell membrane is semi-permeable and allows water to pass through them more rapidly than sugar. The driving force for water removal is the concentration gradient between the solution and intracellular fluid. Osmotic parameters like sugar gain and water loss are correlated with osmosis time. It involves the dehydration of fruit slices in two stages, removal of water using sugar syrup as osmotic agent and subsequent dehydration in the air drier where moisture content is further reduced to about 15% to make the product shelf stable. The quality of osmotically dehydrated product is near to fresh fruit in terms of colour, flavour and texture. Osmotic dehydration of Karpuravalli banana in 60-70% sugar syrup for 12 hours followed by drying in hot air oven at 50°C gave high quality figs with excellent colour and taste. Wrapping the figs in cellophane paper and packaging in polymeric containers gave an extended storage life of 6 months under ambient conditions. In a study undertaken to develop organoleptically acceptable and shelf stable plantain product applying osmotic drying technique, it was found that a sugar concentration of 70° Brix, 60 minutes immersion in osmotic solution and heating to a temperature of 60°C gave organoleptically sound and shelf stable dried plantain product (Ukkuru and George, 1996).

Chavan *et al.* (2010) reported that the osmo-dried banana slices prepared with sulphur fumigation @ 2 g /kg slices for 2 h followed by soaking in 60°Brix sugar syrup containing 0.1% KMS + 0.1% citrate + 0.2% ascorbic acid were found better with respect to colour and appearance, flavour, texture, taste and overall acceptability with non-stickiness of the product. Storage study showed that there was marginal decrease in moisture content and organoleptic quality and increase in TSS, total sugars and reducing sugars content of osmodried banana slices. The products were found microbiologically safe and sensorily acceptable up to 6 months storage at ambient condition.

The effects of temperature and initial sugar syrup concentration on osmotic dehydration of banana slices were studied by Pokharkar and Prasad (1998) at an initial ratio of banana slices: sugar syrup as 1:4. The water loss and sugar gain varied with sugar concentration as well as temperature. The movement of water and sugar was modelled for water loss and sugar gain by the banana slices. The developed model can be used for predicting water loss and sugar gain during osmotic dehydration of banana slices within the range of experimental study.

The effect of duration of heat treatment (blanching) on the firmness, dehydration rate, colour, and pectin esterase (PE) activity, was investigated in sulphited, unsulphited, blanched and non heat-treated intermediate-moisture banana (IMB) by Levi *et al.* (1980). The product was significantly firmer if the bananas had been blanched. Firmness as measured with a texturometer, also increased with increasing SO_2 concentration. The average rate of water removal (dehydration) increased, with increasing blanching time, up to about 4 min, and increased also with increasing SO_2 concentration. There was a decrease in PE activity in both blanched and non-heat-treated IMB, with increasing SO_2 content. This exhibits the effect of sulphiting on pectolytic enzyme activities.

Moisture content is an important attribute which determines the storage life of any dried/dehydrated product. The removal of water in a foodstuff during drying occurs via two mechanisms: migration of water within the foodstuff from inner portions to the surface and evaporation of moisture from the foodstuff surface into the air. The former is considered as the most common moisture migration during drying, and has been used to explain the drying kinetics of banana (Mowlah *et al.*, 1983; Garcia 1988; Sankat *et al.*, 1996). Hot air drying, sun drying and microwave have been employed for producing banana figs. Sun drying is most commonly employed in many developing countries. A technique to rapidly evaluate the moisture content can help in deciding the duration of drying and the quality of the finished product. It was found that ripe banana halves dehydrated to a moisture content of abut 30%, followed by bleaching and / or sulfating, are quite astringent (Ramirez-Martinez *et al.*, 1977).

Krokida and Maroulis (1999) examined the effect of microwave and microwave-vacuum on increased product porosity and colour changes. They showed that microwave drying increases elasticity and decreases viscosity of product. Krokida *et al.* (1998a) studied the effect of freeze-drying conditions on shrinkage and porosity of banana, potato, carrot, and apple. They found that final porosity decreased as sample temperature increased. Krokida *et al.* (1998b) also examined the effect of drying conditions on color change during conventional and vacuum drying of those fruits. Rate of color changes was found to increase as temperature increased and air humidity decreased. Other workers (Robinson 1980) investigated the improvement of banana dehydration and used the results in designing a commercial banana drying plant.

Ripe banana, cut to 10mm thick slabs were osmotically treated in sugar solutions of 35, 50° and 65° Brix for 36h (Sankat *et al.*, 1996). These slabs, with Total Soluble Solids (TSS) contents of 26 °, 34 ° and 39° Brix, respectively, as well as freshly cut but untreated slabs (15° Brix) were air dried in a cabinet type tray drier to near equilibrium conditions at fixed temperatures from 40 to 80°C and at a constant air speed of 0.62m s^{-1}. Increasing the drying air temperature significantly enhanced the drying rate and the K-values, except at 80°C when the rates fell, possibly because of case hardening of the slabs. Reducing the slab thickness also improved the drying rate, but increasing the air speed to 1.03m s^{-1} did not have any profound effect. As the sugar content of the banana slabs increased through the osmotic treatment, drying rates fell. Calculated apparent moisture diffusivities at 60°C ranged from 34.8×10^{-10} m^2 s^{-1} (fresh slab) to 8.8×10^{-10} m^2 s^{-1} for dried (39° Brix) slabs. The moisture diffusivity was significantly lowered as the moisture content dropped with increased levels of sugar. Previously osmosed and then air dried banana slabs showed appealing colour and texture compared to the fresh banana.

15.7 Banana Jam

Jams can be made from every type of fruit. Banana jam is made by cooking ripe fruit pulp with equal quantity of sugar along with pectin and acid in right proportions till it gives a good set. Several varieties of banana are suitable for making jam. This product has good commercial value and is used in several mixed fruit jams too (Narayana and Mustaffa, 2000). Often in a mixed fruit jam, whole fruits or residuals of different types of fruits in a processing factory is made use of. Sometimes over-ripe fruits culled out during processing goes waste. These can be effectively utilized for making jams. The high temperatures of processing makes it safe and addition of colour and essence mask the identity of each components. Jams were prepared from overripe fruits in different combinations and investigated for various characteristics. The use of different combinations of fruit pulps having different levels of sugar (°Brix) : acid ratios affect the yield of jams (Singh *et al.*, 2009). The pulp combinations give differences in natural colors of jams. They could easily be stored for 60 days maintaining good microbial, sensory and textural qualities of jams.

The fruit-flavored yogurts, having probiotic value could be produced from evaporated cow's milk (19.75% dry matter) containing 16% jam prepared with an equal weight of sugar and fruit (sour cherry, orange, strawberry, or banana). It is suggested that these types of yogurt should not be stored longer than 7 days (Con *et al.*, 1996).

15.8 Ready-To-Serve (RTS) Beverage

The conversion of fruits into juice was originally developed as a method for making use of surplus supplies to the fresh market but, while it still fulfills this function, juice production in now firmly established in its own right (Hulmes, 1971).

India produced about 152 thousand tonnes of Ready-to-serve beverages per annum from different fruits such as mango, orange, pineapple, guava and grapes during 2000 (Saravana Kumar and Manimebalai, 2001). The rising number of health-conscious consumers is giving a boost to fruit juices. Indian consumers are shifting from fruit-based drinks to fruit juices as they consider the latter a healthier breakfast/snack option. Within the beverages market, the fruit-based beverages category is one of the fastest growing categories, and has grown at a CAGR of over 30% over the past decade. At present, the Indian packaged juices market is valued at INR 1100 crore (~USD 200 million) and is projected to grow at a CAGR of ~15% over the next three years (http://www. technopak. com/Files/Packaged_Juice_Market_in_India.pdf accessed on 23.11.2013).

There is tremendous scope of adding fruit juices to sweeten aerated water. Besides quenching thirst they would also provide nutrients. Coupled with increasing the demand for soft drinks, there is considerable scope for developing naturally existing nutrients rich fruit juice beverage. The pulp/ juice of mango, pineapple, banana, orange and other fruits are easily digestible, highly refreshing, thirst quenching, appetizing and nutritionally far superior to many synthetic and aerated drinks. Banana juice is useful to combat many diseases viz. anemia, blood pressure, heart diseases, ulcer, seasonal effective disorder, *etc.* Banana is a potential fruit which can capture a notch in the beverage industry as health drink.

However, unlike juicy fruits like grape, orange, lime, etc., banana juice is tightly bound to the pulp and cannot be separated by squeezing. The depectinization of fruit pulp to facilitate juice extraction and clarification is carried out as a routine by adding commercial enzymes (Hulmes, 1971). Sreekantaiah *et al.* (1971) reported liquefaction of banana using pectic enzymes. The processing quality of banana beverage depends on the biochemical composition of fruit and the variety. Through pectolytic enzyme clarification a clear transparent juice to an extent of 75-80% of pulp weight could be obtained with a TSS content of 23-26°Brix depending on the variety. The clarified juice could be dispensed into ready-to-serve beverage by adjusting TSS to 18-20°Brix and acidity to 0.3% with sugar, citric acid and water. After chilling, it forms an excellent thirst quenching and nutritious drink (Narayana *et al.*, 2003; Narayana *et al.*, 2002c).

Narayana *et al.* (2002c and 2003) standardized a process for making RTS banana beverage. Ripe banana cv. Poovan pulp was treated with 0.5% (W/V) pectinase enzyme, incubated at 50°C for 2 hours and juice was extracted. The juice recovery was 62% and it had a TSS content of 21.5° Brix and acidity of 0.83% initially. The extracted juice was dispensed into- RTS beverage with a TSS of 16-17°Brix and acidity of 0.3% by mixing with sugar, citric acid and water followed by pasteurization and bottling. The quality changes during storage upto 90 days showed an increase in TSS, reducing sugars and total sugars while acidity and NEB decreased. No mold or bacterial growth was detected till 90 days of storage.

Koffi *et al.* (1991) reported that two different combinations of pectinase, cellulose, amylase and hemicellulase were more effective in reducing viscosity and improving filterability of both green and ripe banana purees. KMS 100 mg /litre was more effective in producing a light coloured juice with stable colour. Cellulase 0.06% and pectinase 0.05% at 45°C for 2 hours resulted in 73% clear juice yield. Amylase was not effective in the production of banana juice (Pheantaveerat and Anprung, 1993; Sim and Bates, 1994).

The RTS beverage can also be carbonated using carbon dioxide gas at appropriate level. In future, concentrated clarified banana juice will probably be used more as an ingredient in several other fruit juice blends. In fact, first such product launched by Chiquita Brands International Inc. was an orange-banana juice. However, it has grown to become an "any time drink" slightly eroding the market for carbonated beverages (Sole, 1996). Blends of different fruits, vegetables and herbs having high antioxidant potentials are gaining popularity in the health drink segment of beverages.

A study carried out to develop Whey-based Banana Herbal beverage by incorporating *Metha arvensis* extract (0-4%) showed that sensory scores and overall acceptability of beverage with 0-2% Mentha extract was very good. Due to high probiotic value of whey, therapeutic and antibacterial properties of Mentha, the beverage is said to have good nutritive and prophylactic functions (Yadav *et al.*, 2010).

15.9 Banana Beer and Wine

The process of making banana beer and wine is an age old and traditional practice in Africa. In Central and East Africa, the juice from the ripe "beer banana" fruits are consumed either fresh or fermented with low alcohol content having short shelf life (Suzanne Sharrock, 1996).

Making banana beer in East Africa consists of four main stages; a) forced maturation of green banana, b) pressing and mixing, c) recovery of filtered juice and fermentation and d) racking, filtering and cooling (Davies, 1994). Processing of juice from matured green beer bananas for alcoholic beverages is a feature of the banana-based highland cropping systems of six East African countries (Burundi, Kenya, Rwanda, Tanzania, Uganda and Zaire). Thomas (1940) reported that 'Mbide' varieties are true bananas and used principally for making beer or crisps. Narayana *et al.* (2002e) reported the method of production of banana wine using Indian bananas. The method involved that the mashed ripe pulp is mixed with 10% water and after adjusting the TSS to 26°Brix the pulp is sterilized and inoculated with wine yeast, *Saccharomyces cerevisiae* var. *ellipsoideus* and incubated at 24-26°C for 7 days with intermittent aeration. After one week, fermented juice is filtered and kept under anaerobic condition with water seal for secondary fermentation. After two weeks the secondary fermentation is terminated after which the wine is racked or centrifuged, bottled and pasteurized at 50 °C for 20 minutes. The pasteurized wine is kept for ageing, which give wine a characteristic flavour and aroma. Akingbala *et al.* (1994) reported that banana juice and mango juice are fermented to make wine using *Saccharomyces cerevisiae*. Pasteurization of the banana based alcoholic beverages increases the ester and alcohol content.

Beer production consists of mainly four stages: (a) malting (based on germination and treatment of barley); (b) mashing or wort production (extraction and hydrolysis of the components of malt and possibly other cereals, followed by separation of non-soluble components and boiling with hops); (c) fermentation stages (in most cases divided into main fermentation and lagering or maturation); and (d) down-stream processing (filtration, stabilization, bottling, and other treatments of the final product (Linko *et al.*, 1998). Presently, due to improvements in technical, biochemical, microbiological, and genetic inventions in modern brewing industry there are many variations of this brewing process (Bamforth, 2000; Wilaert and Nedovic, 2006; Branyik *et al.*, 2006; Rivera *et al.*, 2009). New techniques and processes have also been developed using several adjuncts for production of different beers such as alcohol-free, low-alcohol, or aromatic beers (van Lersel *et al.*, 1999; Odibo *et al.*, 2002; Agu, 2006). Banana being rich in carbohydrates and minerals with low acidity, can also be a raw material for alcoholic fermentation. Because of the high yields and low production costs of bananas, many countries such as Brazil have been utilizing them widely.

Recent research has shown that banana has potential for application in commercial food processing, due to its functional properties and cheap availability, and has become a viable option as an industrial raw material for various activities (Harish *et al.*, 2008; Zhang *et al.,* 2005; Viviani and Leal, 2007). Research has proved that banana juice as an adjunct in brewing processes has great potential in many tropical banana producing countries. Carvalho *et al.* (2009) conducted a study with an objective to evaluate the performance of wort adjusted with banana juice in different concentrations. Their results showed an increase in ethanol production, with approximately 0.4 g/g ethanol yield and 0.6 g/L h volumetric productivity after 84 h of processing when concentrated wort was used. They concluded that banana can be used as an adjunct in brewing methods, helping in the development of new products as well as in obtaining concentrated worts.

In almost all the banana exporting countries there is considerable 'export rejects' which goes as waste i.e. those which do not meet the quality norms for export (Viquez *et al.*, 1981). These can be utilized for production of wine. In the production of banana wine, pectinolytic enzymes have been used to reduce viscosity and to clarify juice from ripe banana pulp (Jaleel *et al.*, 1979). Jackson and Badrie (2002) investigated the effects of different levels of pectolase enzyme at varying incubation periods (1 h, 4 h, 24 h) at 23 ±1°C on physicochemical, microbiological and sensory qualities of banana wines. They found that addition of enzyme had no effects on percentage wine yield, degrees Brix, percentage alcohol, sulphur dioxide (SO_2), pH or titratable acidity except for colour, volatile acidity and tannin content.

Cheirsilp and Umsakul (2008) treated the banana must with pectinase and á-amylase to hydrolyze pectin and starch prior to its use to produce wine. They noticed a decrease of 55% in the viscosity and a 2.7-fold increase in the amount of extracted juice when incubated at 50°C for 2 hours after treating with 0.05% (w/w) of pectinase followed by treating with 0.05% (w/w) of a-amylase for 3 h. A 15% and 39% increase in total soluble sugars and reducing sugars in extracted juice were achieved, respectively. Enzyme-treated banana must was diluted with four volumes of water and then fermented by yeast to produce banana wine. The clarity of the enzyme-treated banana wine was also four fold higher than that of the control at 25 days of fermentation. The concentrations of total soluble solids, total soluble sugars, and alcohol in the enzyme-treated banana wine and the control had no significant differences. Dilution of banana pulp with equal amount of water and direct fermentation with wine yeast gave stronger banana aroma than the enzyme clarified banana juice (Ranjitha and Narayana, Unpublished data).

Singh *et al.* (2009) diluted pulp (with water) from overripe bananas (Musa robusta) and mixed with sprouted sorghum grains as ingredients and compared with ingredients without sorghum. Their study showed that there was higher utilization of fermentable sugars and carbohydrates in the mix when sorghum was present.

Blending of other fruit juices like pineapple and sapota to fermented banana beverage was also attempted, which showed less preference based on sensory evaluation (Kumar and Mishra, 2010).

15.10 Banana Vinegar

Vinegar is produced from banana wastes in Panama by subjecting to alcoholic and acetic fermentation (Ortega and Blanco, 1986). A simple technique for processing rejects or overripe 'Robusta' banana into vinegar was developed by Suresh and Ethiraj (1991). The pulp was crushed with equal volume of water to which 40mg/kg pectinase was added This gives a clear juice, which was seeded with a strain of *Saccharomyces cerevisiae var. ellipsoideus* to initiate fermentation. A good quality alcoholic base for making vinegar containing 5-6% acetic acid was obtained. Simmonds (1966) reported that vinegar has been prepared by fermenting a mash of banana pulp and peel.

15.11 Banana Catsup

In Philippines, bananas are used to produce ketchup, which is sold commercially. It resembles tomato ketchup in appearance but not in flavour. A recipe for preparation of banana sauce has been developed at National Research Centre for Banana, Trichy, where raw 'Monthan' variety of banana was blanched, peeled, mashed and cooked with spices and vinegar to yield banana sauce with good acceptability. The product is stable for almost a year (Narayana *et al.*, 2002f).

15.12 Banana Inflorescence Relish

Banana inflorescence is used as a vegetable in all the southern and eastern states of India. Several ready to eat products are made using banana inflorescence, notable ones are lentil-banana flower mixed vada, fried vegetable, pakoda and pickle or relish. A recipe was standardized for banana flower relish at National Research Centre for Banana, Trichy based on which several small scale industries have been established (Narayana *et al.*, 2003b). The inflorescence of Mysore banana (*Musa paradisiaca* cv. Mysore) was investigated for its antibacterial and antioxidant activities using various solvent extractions (Padam *et al.*, 2012). Buds and bracts of the inflorescence showed a wide spectrum of inhibition against foodborne pathogenic bacteria such as *Staphylococcus aureus*, *Bacillus cereus, Listeria monocytogenes* and *Vibrio parahaemolyticus*. Methanolic extract of buds proved to have the strongest antibacterial and antioxidant activities. The inflorescence of Cavendish banana varieties are not suitable for consumption due to its bitter taste. The inflorescence of plantains and cooking types are most suitable for making value added products.

15.13 Candied Banana

Candied banana fruits are a kind of sweet meat made using plantain type of bananas. The process involves peeling, making cross cut slices in the pulp and frying of slices in cooking oil. The fried pieces are soaked in sugar syrup of 60°Brix and coated with dried ground sugar (Narayana *et al.*, 2003g).

15.14 Other Banana Based Products

The technology for several other value added products based on banana flour like biscuits, papads, health drink, baby food, sweet chutney, cakes, etc have been developed at NRC for Banana, Trichy (Narayana *et al.*, 2002g).

Dried, ground and roasted green bananas are also used as coffee substitute in some places. Starch can be extracted from banana and plantain pseudostem. Banana starch has been used for producing glue used in manufacturing of cartons for export of fresh banana (Suzanne Sharrock, 1996). Maini (1999) reported that banana pseudostem (20-25 tonnes) from 1000 banana plants when extracted provide 5% edible starch. Dried banana and banana/cassava pancakes called 'Kabalagala' are consumed in Uganda (Aked and Kyamuhangire, 1996). Ogazi (1996) reported that in Nigeria unripe plantain was cooked, the peel and pulp were pounded to obtain a meal and eaten with any of the popular soups. Various delicious banana products (savory and sweets dishes) were prepared by Manimegalai *et al.* (1999) and Anon (2002) using raw and ripe plantain, stem and flower.

Green stems, rootstock or corm, sun dried ripe banana peels provide feed for cattle and sheep. Ripe banana rejects, supplemented with protein, vitamins and minerals are also used for fattening hogs. Good silage can be made from equal parts of chopped green bananas and grass or from chopped green banana mixed with 1.5% molasses. In Philippines meal made from dehydrated rejected bananas could form 14% of total broiler rations without any adverse effect. Peeled leaf sheath are used fresh or after drying as packaging material for flowers, betel leaves, fruits and similar other items. They are stripped into shreds, dried and used for tying packages and making garlands in southern India.

Chapter – 16

Utilization of Field Residues and Processing Waste

The cultivation of bananas is an activity indigenous to the tropics. Banana is a fast growing and high biomass-yielding plant. The volume of waste produced is double the volume of bananas produced. The wastes from banana include the pseudostem, true stem, rhizome, leaves, and peels/skins. Besides these often large quantities of immature fruits end up as waste due to falling of banana plants during windy seasons. Export rejects (fruits unfit for export due to poor quality) is common in many banana exporting countries. Though the tender banana stem is generally used as a vegetable due to its high medicinal values in India, in comparison to the volumes of waste produced, its utilization is miniscule. Banana pseudostem being fibrous is a good source of cellulose, which can be used as raw material for paper making. Banana peels seem to be under utilized as potential growth medium for local yeast strain, despite their rich carbohydrate content and other basic nutrients that can support yeast growth.

Bioethanol is produced from a wide variety of raw materials including corn starch, sugar cane and lignocellulosic waste materials. With a suitable strain, ripe banana peels can be utilized as a substrate for growing yeast and producing bioethanol. The main cost element in bioethanol production is the feedstock and these being waste, could be economical compared to corn or beet.

Large quantities of reject bananas are often available for animal feed, especially in banana exporting countries. Both cattle and pigs relish ripe bananas, but as an animal feed, bananas are mainly used for feeding pigs. A good silage can be made from equal parts of chopped green bananas and grass, or from chopped green bananas mixed with 1.5% molasses and on such a diet, pigs tend to have less carcass fat. Unripe bananas and plantains can also be dried and made into a meal which can be used to substitute up to 70-80% of the grain in pig and dairy diets.

Banana meal has also been used in poultry diets, but high levels in the diet tend to depress growth and reduce feed efficiency. Banana and plantain pseudostems are also fed either fresh, or chopped and ensiled, to cattle and pigs.

The banana industry in Costa Rica, Ecuador and Columbia are one of the largest industries which produce banana by-products like, banana peels, banana stems, and liquid wastes. Currently these wastes are dumped in landfills, rivers, oceans and unregulated dumping grounds. When they reach these destinations, they form huge masses of putrefying wastes that attract insects and scavengers, spread disease, contaminate groundwater, and have emanate foul odors (Leon, 1997). Effective Microorganisms (EM) are currently being used in many countries as a beneficial microbial inoculate for processing organic materials so that they can be recycled back into agricultural systems (Afzal *et al.*, 1994). Experiments were carried out in Costa Rica using EM to process all forms of banana wastes into rich amendments that can improve soil structure and increase crop yields and quality. The products made using wastes from the banana industry can be used as beneficial microbial inoculates for soil regeneration and fertilization (Higa, 1996).

16.1 Banana Alcohol

Waste bananas are those that do not meet export standards due to size variation or blemishes or are the excess resulting from overproduction. Although the common practice of handling waste bananas is to dispose of them as garbage, they represent a potential energy feedstock which may be especially suited for ethanol production. Bananas grown for export are typically harvested and transported to a packing shed for grading. The grading rejects then require disposal. Because they have already been collected and brought to a centralized location, the waste bananas are a low-cost, concentrated biomass feedstock (Hammond *et al.*, 1996).

Laboratory studies were conducted to assess the ethanol production potential from waste bananas. Over a 10-day ripening period, there was a 9% loss of fresh weight by the day 6 and a 15% loss by the day 10. Ethanol yields from normal ripe bananas were: whole fruit-0.091, pulp 0.082, and peel-0.006 l/kg of whole fruit. Ripeness effects on ethanol yield were measured as green-0.090, normal ripe-0.082, and overripe-0.069 l/kg of green whole bananas. Enzymatic hydrolysis was necessary for maximum yields. Diluting with water was not essential for effective fermentation (Hammond *et al.*, 1996).Bananas contain mostly carbohydrates, with only small amounts of fiber. The high nonstructural carbohydrate (NSC) and low fiber contents make bananas a good feedstock for ethanol production. The carbohydrate is nearly all starch when the bananas are harvested green, but rapidly converts to sugar during ripening. The primary sugars in bananas are glucose (19-22% of total sugars), fructose (12-17% of total sugars), and sucrose (62-68% of total sugars). At the same time, there is a loss of fermentable material due to native metabolic activity. By allowing the bananas to ripen before fermentation, enzymatic hydrolysis could possibly be eliminated, leading to a reduction in energy and input costs. The NSC content of the bananas suggests that the fruit could be fermented without diluting with water. Ideally, the sugar content of the unfermented mash should be about 15-2510 (w/w); this will produce about 8-12% ethanol (v/v). Conversion of the structural carbohydrates to energy will probably not be feasible because of the small amounts contained within the fruit.

In Uganda and Sudan, banana beer is distilled to produce banana alcohol or "Waragi". The commercial or medicinal alcohol from banana has been produced in several countries. Large quantities of non-exportable Cavendish bananas in the Philippines were used for producing juice and then converting into alcohol. A technique was described for preparation of inoculum, immobilization of yeast and fermentation of juice to alcohol (Del Rosario and Pamatong, 1985). Tewari *et al.* (1986) reported that banana peel were subjected to comparative saccharification treatments; sulfuric acid, steam under pressure and the enzymatic action of a cellulase. The fermentation was carried out with *Saccharomyces cerevisiae var. ellipsoideus*. The alcohol yield was 150 ml /kg peel. A process for production of commercial grade alcohol from waste peel of ripe banana fruits through saccharification followed by fermentation and distillation was standardized by (Narayana *et al.*, 2007). The yield of alcohol was around 200 ml/ kg peel.

16.2 Banana Pseudostem Fibre and Fibre Reinforced Products

It is well known that plant fibers also known as lignocellulosic fibers (LC) have been one of the attractive fillers for different types of polymers including rubbers as well as for ceramic matrices due to some of their unique characteristics unparalleled with any other reinforcing/filler materials.

Increasing attention has been paid to the use of renewable resources particularly of plant origin keeping in view, the ecological concerns, renewability and compliance of government laws in using such materials. In addition, use of their composites have established comparable performance with those of glass fiber composites with possibility for their use as structural components as well (Bledzki and Gassan, 1999; Burgueno *et al.*, 2004; Holbery and Houston, 2006; Joshi *et al.*, 2004; Netravali and Chabba, 2003). Systematic studies on various lignocellulosic fibers (banana, sugarcane and sponge gourd) including development of composites using these fibers were undertaken by (Guimarães *et al.,* 2006; Pinto, 2007). X-ray diffraction patterns of these three fibers exhibit mainly cellulose type I structure with the crystallinity indices of 39%, 48% and 50% respectively for these fibers. Morphological studies of the fibers revealed different sizes and arrangement of cells. Thermal stability of all the fibers is found to be around 200 °C. Decomposition of both cellulose and hemicelluloses in the fibers takes place at 300 °C and above, while the degradation of fibers takes place above 400 °C. These data may help finding new uses for these fibers (Guimaraes *et al.*, 2009).

The total biomass in banana is reported to be in the range of 54.60 to 96.00 tonnes/ha tonnes/ha and expected fibre yield from the waste parts of banana is in the rage of 36.80 to 483.11 kg/ha (Uma *et al.*, 2003). The scope of utilization of banana fibre in India was reviewed by Ray *et al.* (2012). It is estimated that 60-80 tonnes/ha waste biomass (pseudostem, leaves and rhizome) is generated during the cultivation of banana in India. It is also estimated that annually 17,000 tonnes of fibre can be extracted from the waste portions of banana plant valued approximately INR 850 million (Ray *et al.*, 2012). Banana fibre is more superior in quality in terms of strength, fineness and spinnability compared to bamboo and ramie fibre. In terms of value, it is highly priced than hemp, kenaf and flax at 0.43 to 0.81 US$ per kilogram (Mukhopadhyay *et al.*, 2008). Kulkarni *et al.* (1982) investigated the cellular structure of banana fiber using optical microscopy. It showed that banana fiber consist of four kinds of cells namely xylem, phloem, schlerenchyma, and parenchyma and further they observed how the mechanical property varied in reference to the diameter of the fiber.

Venkateshwaran and Elayaperumal (2010) reviewed the possibilities of use of banana fiber for reinforcement of plastics and composites. The incorporation of banana fiber in cement reinforcement was reported by Savastano *et al.* (2003 & 2005). They suggested that the mechanical properties of these composites in terms of water absorption (<25%) were good and hence is suitable for use as building material.

The incorporation of banana fiber in thermoset plastics was attempted by many workers. Mukherjee and Satyanarayana (1984) showed in their investigation that the banana fiber with polyester composites had the same specific modulus (2.39) as that of glass fiber plastics, which is used in several areas.

Kumar *et al.* (2008) have carried out research on preparing green composites using banana fiber and soy protein as matrix. In their work the alkali treated and untreated fibers were incorporated into soy protein isolate (SPI) with different amounts of glycerol (22-50%) as plasticizer. Their results showed that the alkali treated fiber reinforced with soy protein composites had tensile strength upto 82% and 96.3% respectively compared with soy protein film without fibers. Biodegradability test showed the composites are 100% biodegradable.

Although studies have shown that the cellulose fiber has suitable features to industry, the yield is low because pseudostem has about 90% of water. Bananas and plantains are frequently used as a source of fibre. Banana fibre is extracted from dried petioles and pseudostem of plant, which is used extensively in manufacturing of certain papers, particularly where great strength is required. The paper is used for, amongst other things, making tea bags and bank notes. The fibre has numerous other uses including textile manufacture, making ropes, strings, threads and for production of handicraft as baskets, toys, table mats, wall hangings and lamp shades. Mechanical devices for efficient extraction of fibre from banana pseudostem have been developed by Central Institute for Research on Cotton technology, Mumbai (Nachene and Uma, 2003). The pseudostem fiber has been used mainly in handicrafts.

16.3 Banana Pseudostem Juice

The composition of banana peel revealed that it is rich in fiber and minerals, particularly potassium, calcium and magnesium (Emag, 2007). This information indicated the possibility of presence of these minerals in the sap also. The amount of minerals in coconut water is comparable with that of pseudostem sap. Pseudostem juice is richer in potassium than commercial sports drinks. Heat treatment is a necessary step in processing of pseudostem juice into sport drink either to eliminate bacteria population or to denature PPO enzyme.

The Banana stem is rich in medicinal and curative properties. The central core stem adds as valuable dietary food. Juice from banana stem (banana stem juice) is effective in treating a variety of urinary infections and to dissolve kidney, prostrate and gall bladder stones (Poonguzhali and Chegu, 2004). The concentrated vitamins, minerals and enzymes are responsible for their curative properties (Prasad *et al.*, 1993). Regular use of fresh unpasteurized juices will flush the liver which results in improvement in the hair, skin and nails. Banana stem juice powder was obtained by drying banana stem juice in hot air oven at 70-75 °C for 18-20 hrs. Various physico-chemical constituents of stem juice and its powder were analyzed. Active bio-molecules present in the stem juice was separated by column chromatography for further identification and characterization (Basanth Kumar *et al.*, 2010).

16.4 Banana Starch

Starch is an important naturally occurring polymer with diverse applications in food and polymer science. Annual worldwide starch production is 66.5 million tons (FAOSTAT, 2002) and the modern food industry's growing demand for starches has created interest in identifying new sources of this polysaccharide (Betancur-Ancona *et al.*, 2004). Starch accounts for a significant fraction of a large range of crops. Production of starch from rejected bananas could be a source of income and employment for major banana producing countries. The starch content in banana is comparable with that of endosperm of corn and flesh of sweet potato. Banana contains as much as 70% of starch in some varieties. Fruits such as banana and mango are promising new starch sources since they already form part of the worldwide food system. Current starch research has focused on the search for non-conventional starch sources with diverse physicochemical, structural and functional characteristics that provide them with a broad range of potential industrial uses. Given its properties, square banana starch has potential applications in food systems requiring high temperature processing, such as jellies, sausages, bakery and canned products. It is inappropriate, however, for use in refrigerated or frozen foods (Torre-Gutie´rrez *et al.*, 2008).

Because of the problem of obesity the development of nutraceutical foods with a low glycemic index has become necessary. Cookies with unripe banana flour (UBF) were prepared with very few ingredients in the formulation to avoid fat and digestible carbohydrates. Proximate composition and starch digestibility were carried out in the cookies. Hydrolysis percentage and predicted glycemic index decreased when the UBF increased in the composite that is related with the resistant starch content (Agama-Acevedo *et al.*, 2012). Showing some resistance to human digestive enzymes, the slow release of glucose from RS results in reduced energy intake by the intestinal cells, which is evident from a low glycemic index of the non-digested starch. This can help improve glucose regulation in diabetes and facilitate weight control for the obese. The non-digested starch in the large intestine is fermented by colonic microflora, producing short-chain fatty acids that encourage the growth of beneficial bacteria. This may lead to healthier colon cells and help prevent the development of colon cancer. In addition, a diet high in RS can reduce blood cholesterol and triglyceride levels due to higher excretion rates of cholesterol and bile acids. Overall, increasing the RS content in the diet has the potential to provide several significant health benefits and an added value to food products (Vatanasuchart *et al.*, 2012).

16.5 Banana Waste for Mushroom Production

Cultivation of oyster mushroom is becoming more popular throughout the world because of their ability to grow at a wide range of temperatures utilizing various lignocellulosic material (Khanna and Garcha, 1981). Both pseudo-stem and leaves of the banana tree have high content of lignolitic fibers with high potential for development of edible mushrooms. Ahmed *et al.* (2011) observed that 50% cotton waste + 50% banana leaves was most suitable for the cultivation of Oyster mushroom (*P. ostreatus*).

de Carvalho *et al.* (2012) evaluated the oyster mushroom *Pleurotus ostreatus*-(Jacq.: Fr.) Kumm. cultivation in substrates based on different combinations of wastes of banana (leaf, pseudo-stem and pseudo-stem + leaf) and in different banana cultivars - *Musa* spp. (Thap Maeo, Prata Anã, Pelipita and Caipira). Organic matter loss in the substrate by action of the fungus was also evaluated during that period. It was observed that the pseudo-stem waste provided the best averages of biological efficiency among all cultivars tested and best rates were obtained by Thap Maeo (61.5%). The highest organic matter loss (OML) was obtained from pseudo-stem + leaf wastes (Prata Anã – 78.6%; Thap Maeo – 67.6%; Pelipita – 64.8%; Caipira – 60.6%). Therefore, the use of those wastes showed itself viable for *P. ostreatus* cultivation due to its availability and low cost, besides decreasing discards to environment.

Seven agricultural wastes (saw dust, coir, sugarcane baggase, sugarcane leaves, cotton waste, banana leaves and rice straw) were used as substrates or nutrient source for the production of mushroom (*Pleurotus florida*) to investigate the nutritional composition of mushroom. In some cases significant variation was found in nutritional parameters among the mushrooms grown in different substrates. The amount of protein found in mushroom cultivated in banana leaves was significantly higher than in any other substrate. Also lipid, fiber, carbohydrate and ash content differ although in less extent. In most of the cases nutritional composition of mushroom corresponded to the nutritional composition of substrates (Khan *et al.*, 2008).

16.6 Paper and Pulp from Banana Waste

Paper is made from wood pulp for which forests trees of lesser timber value are used in many parts of the world. However, there is growing importance for the utilization of non-wood plant fibers, as an alternative in the manufacture of pulp, paper and paper board (Darkwa, 1988).Many non wood fibers such as bamboo, jute, straw, rice, and abaca are currently used in small commercial pulping operation. Other agriculture residues such as banana stem posses characteristic suitable for paper making (Franco *et al.*, 1981; Saikia *et al.*, 1997). Other well-known species are abaca and other wild banana plants used as a source of fiber for the paper and cordage industries (Kane and Marathe, 1949; Hiekal and Fadl, 1977). After harvesting the banana fruit, the plant is cut down and thrown away mostly as waste. Banana stem pulping has been the subjected to conflicting reports that showed pulping agent like caustic soda, caustic soda and sodium sulphite or calcium bisulphite do not produce an easily bleachable pulp (Hiekal, 1976). The use of renewable agricultural by products for pulp production as non wood cellulosic fibers would be great advantage for countries with limited wood forests, and would increase the profit of farmers in developing countries. There is an increasing consumption of wood fiber based products, such as panel, composites, textile grade fibers, paper and boards.

Banana stem is a waste material after harvest of fruits and is mostly cut and thrown in the field which disintegrates over a period of time. This waste biomass can be used for manufacturing of pulp for the production of fiber, film and paper. Pulping is done to liberate the fibers from lignin and hemicelluloses, which can be accomplished chemically or mechanically or by combination of both. Chemical pulping is accomplished by the use of chemicals to separate the lignin fraction of lignocelluloses materials from the cellulose. Chemical separation results in little or no effect on the fiber length. Kappa number, yield, viscosity limit index (cm3/gm) is used to describe the extent of lignin removal in the cooking process. Kraft process, sulphite process , soda process , formic acid /acetic acid solvent and Urea/NaOH solvent system are the five pulping techniques available. Heikal *et al.* (1977) compared the kraft and nitric acid pulping of retted and unretted banana chips. They reported that pulping of retted sample gave pulp with better chemical and strength properties than that form unretted samples. The gummy material is removed by acid extraction before kraft pulping (Jacob and Prema, 2008; Jacob *et al.*, 2007). Both the unbleached and bleached pulps possessed satisfactory strength properties suitable for writing and printing paper (Gayatri and Bala, 2006).

The result of pulping of banana pseudo-stem suggest that it can be used as potential source of cellulose to make high valued cellulose based products such as paper, fiber, *etc*. (Kumar and Kumar, 2011).

16.7 Vermicompost and Vermiwash from Banana Waste

After the harvest of the banana fruits the whole plant (leaves, stem and rhizome) is left in the agriculture field for natural degradation, which takes several months. Earthworms have been used in the vermiconversion of urban, industrial and agro-industrial wastes to biofertilizers (Elvira *et al.*., 1998). It is well established that a large number of organic wastes can be ingested by earthworms and egested as peat like materials termed as vermicompost (Sangwan *et al.*, 2008). The utilization of organic residuals reduces production costs and eliminates the need for landfill disposal and incineration. Vermicomposting is an appropriate alternative for the safe, hygienic and cost effective disposal of organic waste. A study was carried out to find the potency of vermicompost using *Musa paradisiaca* (banana peel) waste and *Eudrilus eugeniae* earthworm as it effectively decomposes the waste (Achsah and Lakshmi Prabha, 2013). To analyse the efficiency of vermicompost the physicochemical parameters like pH and the level of macro and micronutrient content namely nitrogen, phosphorous, potassium, iron and copper in the vermicompost has been studied. The enzymes (amylase,cellulase and invertase) and total macronutrients (N, P and K) and micronutrients (Fe and Cu) showed elevated levels in vermicompost than control (raw waste). The efficacy of vermicompost has also been checked and studied on the vegetable plant *Solanum lycopersicum* (tomato). The growth parameters namely root length, shoot length and number of leaves has been studied. Finally, it has been compared with the plants grown using chemical fertilizer (NPK). The study suggested that vermicompost obtained from the degradation of *Musa paradisiaca* (banana peel) waste by *Eudrilus eugeniae* is an effective biofertilizer which would facilitate the uptake of the nutrients by the plants resulting in higher growth and yield.

Vermiwash, a foliar spray, is a liquid fertilizer collected after the passage of water through a column of worm activation. It is a collection of excretory products of earthworms, along with major micronutrients of the soil and soil organic molecules that are useful for plants. This bio-liquid is rich in nutrients and plant growth hormones. Vermiwash seems to possess an inherent property of acting not only as a fertilizer but also as a mild biocide. This vermiwash have enzymes and secretions of earthworms which would stimulate the growth and yield of crops and even develop resistance in crops receiving this spray. Such preparation would certainly have the soluble plant nutrients apart from some organic acids and mucus of earthworm and microbes.

16.8 Animal Feed from Banana Waste

There is an increasing demand for animal feed, thereby competiting in the use of crops for human food. The food grain scarcity is resulting in depletion of food stocks due to which it is becoming uneconomic to continue feeding grain to livestock as generously as in the past. In tropical regions where animal protein for human consumption is scarce, it would be appropriate to make use of every crop by-product as animal feed (Le Dividich *et al.*, 1978).

The option of feeding of bananas, another important tropical crop in terms of volume and energy value, has been relatively neglected. This is largely because bananas are principally a human food, but is also partly attributable to the fact that their value as animal feed has not been adequately considered. It is relevant to note in this context that for every tonne of bananas packed and exported to the fruit markets of the world, about 750 kg are rejected as they are unsuitable for export. This discarded fruit represents 10 to 20 percent of the total crop, the variation depending on how culling is done during packing. Besides these, hurricanes lodge or break many banana plants in fruiting rendering it unfit for export or use. These could serve as alternate animal feeds.

Banana skin is a good source of carbohydrates for all phases of animal life. Mainly it contains appreciable amount of cellulose and hemicellulose (60%) ;with a crude fiber content of a banana skin by 13%.

Several studies have shown that artificial feeding starchy banana peels can improve chicken production as seen from weight gain, feed intake, feed conversion, serum cholesterol levels in the blood, meat, liver, feces, and other digestive organs. Artificially feeding starchy banana peel can also produce broiler meat with low cholesterol levels.

The best way of feeding fresh green banana or plantain fruits is to chop them and sprinkle some salt on the slices since the fruits are very low in the in-organic nutrients. Cattle and pigs relish this material. For ensiling purposes, the chopped green bananas or plantains are preferred to the ripe fruits which lose some of their dry matter and, in particular sugars during ensiling. Similarly, green fruits are more easily dried than ripe fruits which are very difficult to completely dehydrate (http://www.fao.org/docrep/003/ t0554e/ t0554e17.htm accessed on 24.11.2013).

16.9. Banana leaf

Banana leaf is considered to be sacred in India and is used extensively in religious ceremonies like pooja, homa, etc. In hotels and restaurants it is used for serving food. The tender leaves when slightly heated over flame becomes soft and pliable which is used for wrapping food. It is used in temples to give sandal paste and Prasad (holy offerings). Among the Hindus in Kerala, the dead are placed in a full length banana leaf before consigning to the flame, as a symbolism of purification.

Banana leaves are essentially the source of photosynthates during the crop cultivation. It is said that to get good yield the plant should have atleast 14-16 functional leaves at the time of inflorescence emergence. However, banana plant puts forth a leaf every week under Indian conditions. These leaves have many other uses too, because of which specific varieties like Ela Vazhai (in Tamil Nadu) are commercially cultivated for harvesting leaves. Use of banana leaves as plates for serving food is commonly seen in southern India, the Philippines, Indonesia, Malaysia, Sri Lanka and Singapore. It is also used for wrapping food and for cooking (both Indian and Thai cooking). While traditionally it is believed that eating hot food served on banana leaf has certain health benefits, steaming the food in banana leaves impart a subtle flavor to the dish (both vegetarian and non-vegetarian). In the Philippines banana leaf is used as wrappers for frying. In Puerto Rico, Dominican Republic and Mexico the banana leaves are used to steam lamb and pork meat preparation as it imparts a pleasant aroma to the dish. As the leaf matures it becomes brittle and tear, hence it is harvested when still tender. The harvested leaves are bundled in dried banana leaves and tied before transporting to distant markets. The pedicel of banana leaf is used for extraction of natural fibre which has several industrial applications as discussed in earlier paragraphs.

The banana leaves contain several polyphenolic compounds having good antioxidant properties. The surface of banana leaf is waxy (natural protection) and non-adherent. Because of this property banana leaves are recommended for dressing burnt wounds. An important factor in the healing of superficial and moderate partial thickness burn is early and effective coverage with a dressing that protects the wound from trauma and dessication and is non-adherent. Gore and Akolekar (2003) compared banana leaf dressing (BLD) and boiled potato peel bandage (BPPB) in the management of partial thickness burn would in humans and found BLD as an effective and cheapest dressing material for burn patients. As systemic antibiotics do not penetrate eschar, topical therapy with antibacterials were found to be beneficial during the healing phase.

Guenova *et al.* (2013) reported that banana leaves displayed wound-dressing properties that equaled those of petroleum jelly gauze dressings and were tested successfully in a clinical setting in postsurgical patients in Uganda, Africa. It is reported to be an excellent alternative wound dressing, combining the desirable properties of modern wound dressing material with low cost.

One of the significant factors causing wound care pain was that the dressing adhere to the sound bed. Results of clinical trials conducted in Taiwan revealed that for relieving wound pain of certain kinds, banana leaf dressing was superior compared to using boiled potato peel bandage, Acticoat, Mepital or Mefix (Chin, 2006).

Recently its usefulness in treating animal burn wounds was also reported. Banana leaves were used for dressing the thermal burns (ranging from 1st to 3rd degree) in bovines for the first time in India (Choudhary *et al.*, 2011). Protective covering to burn area was provided by banana leaves washed with chlorhexidine solution (2%) and smeared with povidone iodine. Seven out of 34 animals (including one pregnant) recovered in a period of 2 to 4 months. Thus, banana leaf which is a field residue or waste also has several domestic, industrial and medical uses.

Bibliography

Abbas, F.M.A., Saifullah, B.R., Yeoh, S.Y. and Azhar, M.E. 2012. Comparing physico- chemical properties of banana pulp and peel flours prepared from green and ripe fruits. Food Chemistry 129: 312–318.

Abbas, F.M.A., Saifullah, R. and Azhar, M.E. 2009. Assessment of physical properties of ripe banana flour prepared from tow varieties of Cavendish and Dream banana. Int.Food Res. J., 16:183-189.

Abdullah, H., Rohaya, M.A. and Zaipun, M.Z. 1985. Physio-chemical changes during maturation and after ripening of banana (*Musa sapientum* cv. Embun). MARDI Research Bulletin (MYS), 13(3):341-347.

Abe, K. and Watada, A.E. 1991. Ethylene absorbent to maintain quality of lightly processed fruits and vegetables. Journal of Food Science, 56: 1589–1592.

Abera, A.M.K., Gold, C.S. and Kyamanywa, S. 1999. Timing and distribution of attack by the banana weevil (Coleoptera: Curculionidae) in East Africa highland banana (*Musa* spp.). Fla. Entomol. 82:631-641.

Abeywickrama, K., Kularathna, L., Sarananda, K. and Abeygunawardena, D. 2004. *Cymbopogon citratus* (lemongrass) and citral a+b spray treatments alone or in combination with sodium bicarbonate in controlling crown rot in embul banana (*Musa acuminata* AAB). Tropical Agricultural Research and Extention 7:104-111.

Abeywickrama, K., Wijerathna, C., Herath, H. and Saranandada, K.H. 2009. An integrated treatment of basil oil (*Ocimum basilicum*) and alum with modified atmosphere to control crown rot disease in Embul banana. Tropical Agricultural Research and Extension, 12(1): 23-30.

Achsah, R.S. and Lakshmi Prabha, M. 2013. Potential of vermicompost produced from banana waste(*Musa paradisiaca*) on the growth parameters of *Solanum lycopersicum*. International Journal of ChemTech Research, 5(5):2141-2153.

Adeniji, T.A. and Tenkouano, A. 2007. Effect of processing on micronutrient content of chips produced from sme plantain and banana hybrids, Fruits 62:345-352.

Adônis Moreira, Pereira, José Clério Rezende, Freitas, Alfredo Ribeiro De. 2009.Nitrogen and Potassium on Yield and Quality of Banana Cultivar Thap Maeo Bragantia, 68(2): 483-491.

Afzal, S., R. Ahmad, T. Hussain and G. Jilani. 1994. Developments for EM-technology to replace chemical fertilizers in Pakistan. In Third Conference on Effective Micro-organisms (EM). Kyusei Nature Farming Center. Saraburi, Thailand. pp. 45-66.

Agama-Acevedo Edith, Islas-Hernández José J., Pacheco-Vargas Glenda, Osorio-Díaz Perla, Bello-Pérez Luis Arturo. 2012. Starch digestibility and glycemic index of cookies partially substituted with unripe banana flour. LWT - Food Science and Technology, 46:177-182.

Agarvante, J.V., Matsui, T. and Kitagawa, H. 1990. Starch breakdown in ethylene-treated and ethanol-treated bananas: changes in phosphorylase and invertase activities during ripening. J. Jpn. Soc. Food Sci. Technol. 37:911–5.

Agu, R.C. 2006. Technical Quarterly—Master Brewers Association of the Americas, 43:277:280.

Aguayo, E., Escalona, V. H., and Artés, F. 2008. Effect of hot water treatment and various calcium salts on quality of fresh-cut´Amarillo´ melon. Postharvest Biology and Technology, 47:397–406.

Ahmad S, Pervez, M.A. and Thompson, A.K. 2007. Effect of harvest maturity and position of hands on the ripening behavior and quality of banana fruit. 127p.

Ahmad Waqas, Iqbal Javaid, Salim Muhammad, Ahmad Iftikhar, Sarwar Aqeel Muhammad, Shehzad Asif Muhammad and Rafiq Awais Muhammad. 2011. Performance of Oyster Mushroom (Plerotus ostreatus) on cotton waste amended with maize and banana leaves. Pakistan Journal of Nutrition, 10(6):509-513.

Ahvenainen, R. 2000. Ready-to-use fruits and vegetables, Flair-Flow Europe Technical Manual. 2000.

Aikman, D.P. 1989. Potential increase in photosynthetic efficiency from the redistribution of solar radiation in a crop. J. Exp. Bot. 40:855–864.

Aked, J. and Kyamuhangire. 1996. Post harvest aspects if highland bananas in Uganda.

Akingbala, J.O., Oguntimein, G.B., Olunlade, B.A. and Aina, J.O. 1994. Effects of pasteurization and packaging on properties of wine from overripe mango and banana juices. Tropical Science. 34(4):345-352.

Akubor, P.I., Adamolekun, F.O., Oba, C.O., Obari, H. and Abudu, I.O. 2003. Chemical composition and functional properties of cowpea and plantain flour blends for cookie production. Plant Foods Hum. Nutr. 58(3):1-9.

Alabadan, B.A., Sunmonu, M.O. and Sunmonu, T.O. 2008. Effect of packaging materials on some nutritional qualities of fruit and vegetables stored in a passive evaporative cooler. Asian. J. Food Ag-Ind., 1(03):184-196.

Ali, A., M. Maqbool., P.G. Alderson and N. Zahid. 2012. Biodegradable novel edible coatings to control postharvest anthracnose and quality of fresh horticultural produce. Acta Horticulturae, 945: 39-44.

Alvindia, G. Dionisio. 2012. Revisiting hot water treatments in controlling crown rot of banana cv. Bungulan, Crop Protection, 33:59-64.

Anjanasree, K.N. and Bansal, K.C. 2003. Isolation and characterization of ripening-related expans in cDNA from tomato. J. Plant Biochem. Biotech. 12: 31-35.

Anonymous. 1988. The Philippines recommends for banana. PCARRD Technical Bulletin Series, 66:26-28.

Anonymous. 2009a. Final Project Report (RPF-III). Studies on handling, storage and processing of banana. National Research Centre for Banana (ICAR), Trichy, Tamil Nadu. India.

Anonymous. 2009b. Final Project Report (RPF-III). Standardization of storage conditions of banana. National Research Centre for Banana (ICAR), Trichy, Tamil Nadu. India.

Aparicio-Saguila'n, A., Sa'yago-Ayerdi, S.G., Vargas-Torres, A., Tovar, J., Ascencio-Otero, T. E., and Bello-Pe'rez, L.A. (2007). Slowly digestible cookies prepared from resistant starch-rich lintnerized banana starch. Journal of Food Composition and Analysis, 20:175-181.

Apte, P.V. and Laloraya, M.M. 1982. Inhibitory action of phenolic compounds on abscisic acid-induced abscission, J. Exp. Bot. 33(1982):826–830.

Ara, N., Basher, M.K. and Hossain, M.F. 2011. Growth, yield and quality of banana (Musa sapientum L.) influenced by different banana varieties / lines and planting time, Tropical Agricultural Research & Extension, 14(2):45-51.

Aradhya, S.M. and Eipeson, W.E. 1999. Technology for Export of Banana by Sea. Proceedings of National seminar on 'Banana Production, Postharvest Technology and Export', held on 15-16th October 1999 at NIPHT, Pune.

Arts, I.C.W, van de Putte B, Hollman PCH. 2000. Catechin contents of foods commonly consumed in The Netherlands. 1. Fruits, vegetables, staple foods and processed foods. J Agric Food Chem 48:1748–51.

Arts, I.C.W., Hollman, P.C.H. 2005. Polyphenols and disease risk in epidemiologic studies. Am J Clin Nutr 81(Supp): 317–325.

Arunachalam, A., Ramasamy, N. and Muthukrishnan, C.R. 1976. Studies on the nutrient concentration in leaf tissue and fruit yield with nitrogen level for Cavendish cones. Progressive Hort., 8:13-22.

Asif, M.H. and Nath, P. 2005. Expression of multiple forms of polygalacturonase gene during ripening in banana fruit. Plant Physiol Biochem, 43(2):177-84.

Athani, S.I., Revanappa and Dharmatti, P.R. 2009. Effect of plant density on growth and yield in banana. Karnataka J.Agric.Sce., 22(1):143-146.

Aurore, G., Parfait, B. and Fahrasmane, L. 2009. Bananas, raw materials for making processed food products. Trends in Food Science and Technology 20:78-91.

Ayub, R.A., Lopes, L.C., Silva, F.C.C-da; Miranda, L.C.G.-de and Conde, A.R. 1996. Determination of harvesting stage of banana cv. Prata (Musa AAB subgroup–Prata) by chemical analysis. Revista-Ceres, 43:247.

Babbar N, Oberoi, H.S., Uppal, D.S. and Patil, R.T. 2011. Total phenolic content and antioxidant capacity of extracts obtained from six important fruit residues. Food Res Intl. 44:391-396.

Badia, I.A. 1987. Processing plantain chips in Honduras. Proceedings of third meeting of International cooperation for effective plantain and banana research. 1987:162-163.

Baldwin E.A, Nisperos, M.O., Chen, X. and Hagenmaier, R.D. 1996. Improving storage life of cut apple and potato with edible coating. Postharvest Biology and Technology, 9:151-163.

Baldwin, E.A., Nisperos-Carriedo, M.O. and Baker, R.A. 1995. Use of edible coatings to preserve quality of lightly (and slightly) processed products. Critical Reviews in Food Science and Nutrition, 35:509-524.

Baldwin, E.A., Burns, J.K., Kazokas, W., Hagenmaier, R.D., Bender, R.J. and Pesis, E. 1999. Effect of two edible coatings with different permeability characteristics on mango (*Mangifera indica* L.) ripening during storage. Postharvest Biology and Technology 17 (3): 215-226.

Baldwin, E.A., Nisperos-Carriedo, M.O. and Baker, R.A. 1995. Edible coatings for lightly processed fruits and vegetables. HortScience 30(1):35-38.

Bamforth, C.W. 2000. Journal of the Science of Food and Agriculture, 80:1371–1378.

Banks, N.H. 1985. Response of banana fruit to Pro-long coating at different times relative to the initiation of ripening. Scientia Horticulturae, 26(2):149-157.

Barkai-Golan, R. and Phillips, D.J. 1991. Postharvest heat treatments of fresh fruits and vegetables for decay control. Plant Dis. 75:1085-1089.

Barry, Cornelius S. and Giovannoni, James J. 2007. Ethylene and Fruit Ripening. J Plant Growth Regul, 26:143-159.

Basanth Kumar, B.S., Narayana, C.K. and Jeyabaskaran, K.J. 2010. Studies on banana stem juice and stem juice powder by tray drying process. Proceedings of Global Conference on "Meeting the Challenges in Banana and Plantain for Emerging Biotic and Abiotic Stresses", held on 10-13th Dec 2010, at Trichy, Tamil Nadu.

Beena Natesh, B., Arbindakshan, M. and Valsalakumar, R.K.1993. Effect of split application of fertilizers in banana Musa AAB 'Nendran'. South Indian Hort. 41: 67-73.

Betancur-Ancona, D., Gallegos-Tintore′, S. and Chel-Guerrero, L. 2004. Wet-fractionation of Phaseolus lunatus seeds: Partial characterization of starch and protein. Journal of the Science of Food and Agriculture, 84:1193-1201.

Bhalerao Amol, Sharma J.P., Rao D.U.M., Singh Premlata and Lal Krishan. 2010. Assessment of Extent of Value Addition in Banana. Journal of Community Mobilization and Sustainable Development 5(1):081-085.

Bhargava, B.S., Singh, H.P. and Chadha, L., 1993, Role of potassium in development of fruit quality. In: Advances in Horticulture, 2, Fruit Crops: Part 2. Eds. Chadha, K.L. and Pareek, O.P.

Bhaskar, J.J., Shobha, M.S., Sambaiah, K. and Salimath, P.V. 2011. Beneficial effects of banana (Musa sp. var. elakki bale) flower and pseudostem on hyperglycemia and advanced glycation end-products (AGEs) in streptozotocin-induced diabetic rats. J Physiol Biochem. 2011 Apr 8. Apr 8. [Epub ahead of print]. PMID:21476022.

Bico, S.L.S., Raposo, M.F.J., Morais, R.M.S.C. and Morais, A.M.M.B. 2009. Combined effects of chemical dip and/or carrageenan coating and/or controlled atmosphere on quality of fresh-cut banana. Food Control, 20(5):508–514.

Bjorck, I.M., Granfeldt, Y., Liljeberg, H., Tovar, J. and Asp, N., 1994. Food properties affecting the digestion and absorption of carbohydrates. American Journal of Clinical Nutrition 59:699S–705S.

Blackbourn, H., Jeger, M. and John, P. 1990. Inhibition of degreening in the peel of bananas ripened at tropical temperatures. III. Changes in plastid ultrastructure and chlorophyll–protein complexes accompanying ripening in bananas and plantains. Ann. Appl. Biol., 117:147-161.

Blankenship, S.M. and Herdeman, R.W. 1995. High Relative Humidity after Ethylene Gassing is Important to Banana Fruit Quality. HortTechnology, 5:94-187.

Bledzki, A.K. and Gassan, J., 1999. Composites reinforced with cellulose based fibers. Prog. Polym. Sci. 24(2):221-274.

Bleecker, A.B., and Kende, H. 2000. Ethylene: A gaseous signal molecule in plants. Annu Rev Cell Dev Biol 16:1–18.

Bonnet Christelle Brunot, Hubert Olivier, Mbeguie Didier Mbeguie, A., Pallet Dominique, Hiol Able, Reynes Max and Poucheret Patrick. 2013. Effect of physiological harvest stages on the composition of bioactive compounds in Cavendish bananas. J Zhejiang Univ-Sci B (Biomed & Biotechnol) Cavendish bananas. 14(4):270-278.

Borthakur, N.N. and Arnold, N.P. 1990. Chemical evaluation of Musa 'Bhimkol' as a baby food. J.Sci.Food Agric., 53:497-504.

Brányik, T., Silva, D. P., Vicente, A. A., Lehnert, R., Almeida e Silva, J. B., Dostálek, P., et al. 2006. Journal of Industrial Microbiology & Biotechnology, 33:1010–1018.

Brecht, J. K. 1999. Postharvest quality and safety in fresh-cut vegetables and fruits. Cooperative Regional Research Project S-294.

Bresnahan, A.Kara., Arscott, A.Sara., Khanna, Harjeet, Arinaitwe, Geofrey, Dale James, Tushemereirwe Wilberforce, Mondloch Stephanie, Tanumihardjo P.Jacob, De Moura Fabiana, F and Tanumihardjo A.Sherry. 2012. Journal of Nutrition 142(12): 2097-2104.

Brett, C. and Waldron, K.W., 1996. Physiology and Biochemistry of Plant Cell Walls. Chapman and Hall, London, UK, p. 256.

Brown, D.J. 1981 The effects of low 02 atmospheres and ethylene and CO_2 production, and 1-aminocyclopropane-l-carboxylic acid concentration in banana fruits. MS thesis, University of Maryland, College Park.

Brummell, D.A., Dal Cin, V., Crisosto, C.H. and Labavitch, J.M. 2004. Cell wall metabolism during maturation, ripening and senescence of peach fruit. Journal of Experimental Botany 55:2029–2039.

Bugaud, C., Daribo, M.O., Dubois, C. 2007. Climatic conditions affect the texture and colour of Cavendish bananas (Grande Naine cultivar). Sci. Hort. 113(3):238-243.

Burdon, J.N., Moore, K.G. and Wainwright, H.1993. The peel of plantain and cooking banana fruits. Ann. Appl. Biol. 123(2):391-402.

Burgueno, R., Quagliata, M.J., Mohanty, A.K., Mehta, G., Drzal, L.T. and Misra, M. 2004. Load-bearing natural fiber composite cellular beams and panels. Composite: Part A 35(6): 645–656.

Butler, A.F. 1960. Fertilizer experiments with the Gros Michel banana. Trop. Agric., Trinidad, 37:31-50.

Cano, M.P., B de Ancos, Lobo, M.G. and Santos, M.1997. Improvement of frozen banana (Musa cavendishii, cv.Enana) color by blanching: relationship between browning, phenols and polyphenol oxidases and peroxidase activities. Z.Lebensm Unters Forsch A.204: 60-65.

Carpita, N.C. and Gibeaut, D.M. 1993. Structural models of primary cell walls in flowering plants: consistency of molecular structure with the physical properties of the walls during growth. Plant J, 3(1): 1-30.

Carreel, F. 1994. Etude de la diversite des bananiers (genre Musa) a l aides des marqueurs RFLP. Thee de la Institut National Agronomique, Paris-Grignon, Pp.90.

Carvalho B.M. Giovani, Silva P. Daniel., Bento, V. Camila., Vicente, A. Antonio., Teixeira, A. Jose., Maria V.Gracas A.Felipe and Almeida e Silva B.Joao. 2009. Bananas as adjunct in beer production: Applicability and Performance of Fermentative Parameters. Appl Biochem Biotechnol, 155:356-365.

Caussiol, L. 2001. Postharvest quality of conventionally and organically grown banana fruit. An M. Sc. Thesis Presented to Cranfield University, Silsoe. pp 160.

Cellini, L., Di Campli, E., Masulli, M., De Bartolomeo, S. and Allocati, N. (1996). Inhibition of Helicobacter pylori by garlic extract (Allium sativum). FEMS Immunology and Medical Microbiology, 13:273–277.

Chang, P.Y., Shuji, F., Ashrafuzzaman, M.D., Naoko, N., Nobuyuki, H. 2000. Purification and Characterization of Polyphenol Oxidase from Banana (*Musa sapientum* L.) Pulp. J. Agric. Food Chem. 48:2732-2735.

Charpentier, J.M. and Martin-Prevel, P. 1965. Culture sur milieu artificial. Carences attenuees ou temporaries en elements majeur, carence en oligo-elements chez le bananier. Fruits, 20:521-557.

Chattopadyay, P.K., Chattopadhyay, S., Maiti, S.C and Bose, T.K.1980.Effect of plant density on growth, yield and quality of banana. In: National Symposium on Banana Production Technology, pp.381-386.

Chaudhuri, P. and Baruah, K. 2010. Studies on plant density in banana cv.Jahaji (AAA). Indian Journal of Hill Farming 23(2):31-38.

Chavan U.D., Prabhukhanolkar, A.E. and Pawar, V.D. 2010. Preparation of osmotic dehydrated ripe banana slices. J Food Sci Technol, 47(4):380–386.

Cheah, L.H., Page, B.B.C. and Shepherd, R. 1997. Chitosan coating for inhibition of sclerotinia rot of carrots. New Zealand Journal of Crop and Horticultural Science, 25(1): 89-92.

Cheirsilp, B. and Umsakul, K. 2008. Processing of Banana-based Wine Product Using Pectinase and A-amylase, Journal of Food Process Engineering, 31(1):78-90.

Chellappan, K.1983. Effects of 2,4-D and GA3 on fruit development and postharvest physiology in bananas. Ph.D Thesis, Faculty of Horticulture, TNAU, Coimbatore.

Chen, H., 1993. Analysis on the acoustic impulse resonance of apples for nondesctructive estimation of fruit quality. Thesis Nr. 236, Faculty of Agricultural and Applied Biological Science, K.U. Leuven, p. 165.

Chervin, C. and Boisseau, P.P. 1994. Quality maintenance of 'ready-to-eat' shredded carrots by gamma irradiation. Journal of Food Science, 59:359-361.

Chillet, M., Hubert, O. and De Lapeyre De Bellaire, L. 2010. Postharvest Disease: Effects of the Physiological Age of Bananas (*Musa*spp.) on Their Susceptibility to Wound Anthracnose Due to *Colletotrichum musae.* Proc. IC on Banana & Plantain in Africa Eds.: T. Dubois et al. ISHS 2010, Acta Hort. 879:419-424.

Choo Chong Li, and Aziz Noor Aziah Abdul. 2010. Effects of banana flour and b-glucan on the nutritional and sensory evaluation of noodles, Food Chemistry 119:34-40.

Choudhury, H., Chandra, K. and Baruah, K. 1996. Variation in total chlorophyll content and its partitioning in Dwarf Cavendish bananas as influenced by bunch cover treatments. Crop Res., 11(2):232-238.

Christophe, Bugaud, Daribo Marie-Odette, Beaute Marie-Pierre, Telle Nelly and Dubois Cecile, 2009. Relative importance of location and period of banana bunch growth in carbohydrate content and mineral composition. Fruits, 64(2):63-74.

Chudawat, B.S., Dave, S.K. and Patel, N.L. 1983a. Effect on the dates and growth of Basrai banana (Musa paradise cv.Basrai). Indian J.Hort., 41:65-68.

Chundawat, B.S., Dave, S.K. and Patel, N.L. 1983b. Effect of close planting on the yield and quality of Lacatan banana. Indian J. Agric. Sci. 53:470-72.

Chundawat, B.S., Dave, S.K. and Patel, N.L.1982. High density plantation in relation to yield and quality in Basrai banana. South Indian Hort., 30:175-177.

Clavijo, H. and Maner, J. H. 1974. The Use of Waste Bananas for Swine Feed: Series EE-No. 6. CIAT (Centro International de Agricultura Tropical), Cali, Columbia.

Con, A.H., Cakmakci, S., Caglar, A. and Gokalp, H.Y. 1996. Effects of different fruits and storage periods on microbiological qualities of fruit-flavored yogurt produced in Turkey. Journal of Food Protection, 4:402-406.

Cordeiro Z.J.M. and Matos A.P.D. 2005. Diseases of banana, Doencas da banana. Informe Agropecuario, 26:12-16.

Cordenunsi, B.R., Lajolo, F.M., 1995. Starch breakdown during banana ripening: sucrose synthase and sucrose phosphate synthetase. J. Agric. Food Chem., 43(2): 347-351.

Crane JH, Carlos FB, Ian M 2005. Banana growing in the Florida home landscape. Institute of Food and Agricultural Sciences (IFAS), University of Florida, Florida.

Croucher, H.H. and Mitchell, W.K. 1940. Use of fertilizers on bananas. Bull. Dept. Sci.Agric. Jamaica.19p

Dadzie BK, Orchard JE 1997. Routin postharvest screening of banana/plantain hybrids criteria and methods. International network for banana and plantain (Inibap), Technical Guidelines. Rome, Italy. pp 75-76.

Dalal, V.B., Thomas, P., Nagaraju, N., Shah, G.R. and Amla, B.L. 1970. Effect of wax coating on bananas of varying maturity. Indian Food Packer, 24:36-40.

Dandamrongrak Rak, Mason Richard and Young Gordon. 2003. The effect of pretreatments on the drying rate and quality of dried bananas. International Journal of Food Science and Technology, 38(8):877-882.

Daniells, J., Jenny, C., Karamura, D. and Tompkepe, T. 2001. Musalogue: A Catalogue of Musa Germplasm. Diversity in the Genus Musa (E. Arnaud and S. Sharrock, comp.) INIBAP, Montpellier, France. pp. 213

Darkwa. N.A.1988. Pulping characteristics of some plantain (*Musa paradisiacal L*) psedostems, Proc. International Non-wood fiber pulp and paper Congress- Pp.973.

Davey, M.W., Keulemans, J and Swennen, R. 2006. Methods for the efficient quantification of fruit provitamin A contents. Journal of Chromatography A 1136(2): 176-184.

Davey, M.W., Mellidou, I and Keulemans, W. 2009a. Considerations to prevent the breakdown and loss of fruit carotenoids during extraction and analysis in Musa. J. Chromatogr. A 1216:5759-5762.

Davey, M.W., Saeys, W., Hof, E., Ramon, H., Swennen, R.L. and Keulemans, J. 2009b. Application of visible and near-infrared reflectance spectroscopy (vis/NIRS) to determine carotenoid contents in banana (*Musa* spp.) fruit pulp. J. Agric. Food Chem. 57:1742-1751.

Davies, G. 1994. The brewing of banana beer in Uganda. Comparison of two artisanal methods. Echos du Cota. 3(64):3-9.

Davila J: Evaluacion de la actividad hemoaglutinante de extractos de plantas de la familia Musaceae [Thesis], Facultad de Farmacia, Escuela de Farmacia, Instituto de Investigaciones, Seccion de Biotecnologia, Universidad de Los Andes, Merida, Venezuela, 2002.

de Carvalho, C.S.M., de Aguiar, L.V.B., Campos, C.S., de Almeida Minhoni, M.T and de Andrade, M.C.N. 2012. Applicability of the use of waste from different banana cultivars for the cultivation of the Oyster Mushroom. Brazilian Journal of Microbiology pp 819-826.

De Costa, D.M. and Erabaduputiya, H.R.U.T. 2005. An integrated method to control post harvest diseases of banana using a member of the Kurkholderia cepacia complex. Postharvest Biology and Technology, 36: 31-39.

De Swardt, G.H. and Maxie, E.C. 1967. Pectin methylesterase in theripening banana. S Afr J Agri, Sci. 10: 501-506.

Debeaufort, F., J.A. Quezada Gallo, and A. Voilley. 1998. Edible films and coatings: Tomorrow's packagings: A review. Critical Rev. Food Sci. 38:299–313.

Debnath, U., Suresh, C.P. and Hasan, M.A. 2001. Bunch management for profitable production in winter developing banana bunches. Indian J. Hort., 58(3): 202-207.

Del Rosario, E.J. and Pamatong, F.V. 1985. Continuous Flow fermentation of banana fruit pulp sugar into ethanol by carrageenan-immobilised yeast. Biotechnology Letters, 7 (11): 819-820.

Del Verde-Mendez, C.M., Forster, M.P., Rodriguez-Delgado, M.A., Rodriguez-Rodriguez, E.M., Diaz-Romero, C. 2003. Content of free phenolic compounds in banana from Tenerife (Canary Islands) and Ecuador. Eur Food Res Technol, 217:287–90.

Desai, B.B. and Deshpande, P.B. 1978. Effects of stage of maturity on some physical and biochemical constituents and enzyme activities of banana. Mysore J. Agric. Sci., 12:193-201.

Desai, B.B. and Deshpande, P.B. 1975. Chemical transformations in three varieties of banana (*Musa paradisiaca).* L) fruits stored at 20 °C. Mysore Journal of Agricultural Sciences, 9(4): 634-643.

Desai, B.B. and Despande, P.B. 1978. Effects of stage of maturity on some physical and biochemical constitutents and enzyme activities of banana. Mysore H.Agric.Sci., 12:193-201.

Dhua, R.S., Ghosh, S.K. and Sen, S. 1992. Standardization of harvest maturity of banana cv.Gaint Governor in West Bengal. In: Advances in Horticulture and Forestry (Ed. Singh, S.P) Vol. 2, Scientific Publishers, Jodhpur. (2):24-35.

Diaz, A., Totte, A.,Giroux, F., Reynes, M. and Raoult Wack, Q.L. 1996. Deep fat frying of plantain (*Musa paradisiaca.* L). I. Characterization of control parameters. Food Science and Technology. 29(5-6): 489-497.

Dinesh Kumar, Pandey, V. and Anjaneyulu, K. 2008. Effect of planting density and nutrient management on growth, yield and quality of micro-propagated banana Rasthali Pathkapoora (AAB). India J. Hort., 65(3):272-276.

Dionisio, G.A. 2012. Revisiting hot water treatments in controlling crown rot of banana cv. Buñgulan. Crop Prot. 33:59-64.

Dominguez-Puigjaner E., Vendrell, M. and Ludevid, M.D. 1992. Differential Protein Accumulation in Banana Fruit during Ripening. Plant Physiol. 98(1):157-62.

Dorais, M., Gosselin, A., 2002. Physiological response of greenhouse vegetable crops to supplemental lighting. Acta Hortic. 580:59–67.

Doshi, J.S., Sutar, R.F. 2010. Studies on pre-cooling and storage of banana for extension of shelf life. Journal of Agricultural Engineering, 47(2): 14-19.

Droby, S., Hofstein, R.,Wilson, C.L.,Wisniewski, M., Fridlender, B., Cohen, L.,Weiss, B., Daus, A., Timar, D., Chalutz, E., 1993. Pilot testing of *Pichia guilliermondii*: a biocontrol agent of postharvest diseases of citrus fruit. Biol. Control 3:47–52.

Eckert, J.W., Ogawa, J.M., 1985. The chemical control of postharvest diseases: subtropical and tropical fruits. Annu. Rev. Phytopathol, 23:421–454.

El Ghaouth A., Arul J., Ponnampalam, R. and Boulet, M. 1992. Chitosan coating to extend the storage life of tomatoes. Journal of Horticultural Science 27: 1016- 1018.

El Ghaouth, A., Arul, J., Ponnampalam, R. and Boulet, M. 1991. Chitosan coating effect on storability and quality of fresh strawberries. Journal of Food Science 56 (6): 1618-1620.

El-Ghaouth, A.,Wilson, C.L., 1995. Biologically-based technologies for the control of postharvest diseases. Postharvest News Inform.6, 5N–11N.

Elsiddig, E.A.M, Elamin, O.M., and Elkashif, M.E. 2009. Effects of plant spacing on yield and fruit quality of some Cavendish banana (Musa spp.) clones. Sudan Journal of Agricultural Research, 14:53-60.

Elvira, C., Sampeelro, L., Benitez, E. and Nagales, R. 1998. 'Vermicomposting of sludges from paper mill and dairy industries with *Eisenia andrei*: a plot scale study', Bioresour. Technol., 62:205-211.

Elzayat, H.E. 1996. Influence of plastic wrapping on storage and quality of banana. Bulletin o Faculty of Agriculture University of Cairo. 47(2): 295-303.

Emag, T. H. 2007. Effects of the Stage of Maturation and Varieties on the Chemical Composition of Banana and Plantain Peels. Food Chem., 103(2):590-600.

Englyst, H.N., Kingman, S.M. and Cummings, J.H. 1992. Classification and measurement of nutritionally important starch fractions. European Journal of Clinical Nutrition, 46(2): S33–S50.

Esguerra, E.B., Hilario, D.C.R. and Absulio, W.L. 2009. Control of Finger Drop in 'Latundan' Banana (Musa acuminata AA Group) with Preharvest Calcium Spray. Acta Horticulturae No.: 837.

Evans Edward and Ballen Fredy. 2012. Banana Market. Published by Food and Resource Economics Department, Florida Cooperative Extension Service, Institute of Food and Agricultural Sciences, University of Florida, Gainesville, FL 32611, published in 2012.

Facundo, Heliofabia Virginia de Vasconcelos; Deborah dos Santos Garruti; Carlos Tadeu dos Santos Dias; Beatriz Rosana Cordenunsi, and Franco Maria Lajolo. 2012. Influence of different banana cultivars on volatile compounds during ripening in cold storage. Food Res. Int., 49: 626–633.

Faisant, N., Gallant, D. J., Bouchet, B. and Champ, M. 1995. Banana starch breakdown in the human small intestine studied by electron microscopy. European Journal of Clinical Nutrition, 49: 98–104.

Faller, A.L.K., Fialho, E. 2010. Polyphenol content and antioxidant capacity in organic and conventional plant foods. J Food Compos Anal 23: 561–8.

FAO, 1981. Food Loss Prevention in Perishable Crops., FAO Agricultural Service Bulletin No.43. Food and Agricultural Organization, Rome, 1981.

FAO, 1990. Diet, Nutrition and the Prevention of Chronic Diseases. A Report of Joint WHO/ FAO Consultation, FAO Corporate Document Repository, FAO, Rome.

FAOSTAT. 2002. Database FAO. Food and Agriculture Organisation of the United Nations. Rome, Italy, 19.2.

Feinberg, B., Olsen, R.L., Mullins, W.R. 1987. Prepeeled potatoes. In: W.F. Talburt and Smith, O. (Eds), Potato Processing. AVI Publishing Co. Inc. 1987. p. 697.

Feriotti, D.G. and Iguti, A.M. 2011. Porposal for use of pseudostem from banana tree (*Musa cavendish*). Downloaded from *www.icef11.org/content/papers/few/FEW893.pdf* on 24/ 7/2012.

FICCI. 2010. Bottlenecks in Indian Food Processing Industry. FICCI Survey- 2010 on Challenges in Food Processing Sector.

Finney, E.E., Ben-Gere, I. and Massie, D.R. 1967. An objective evaluation of changes in firmness of ripening banana using sonic technique. Journal of Food Science. 32:642-647.

Fischer, R.L. and Bennett, A. B. 1991. Role of cell wall hydrolases in fruit ripening. Annu. Rev. Plant Physiol. Plant Mol. Biol. 42:675–703.

Floros, J.D. 1993.The shelf-life of fruits and vegetables. *In:* shelf life Studies of Food and Beverages, G. Charahmbous (Ed), Elsevier Science Publishers, London. 195-216.

Forsyth, W.G.C. 1980. Banana and plantain. *In:* S. Nagy and P.E. Shaw (Eds.). Tropical and subtropical fruits. AVI Publ. Co., Westport, CT, pp. 258-278.

Franco, P.T., et al. 1981. Abaca pulp and paper industry in the Philippines , Proc TAPPI Pulping Conference, pp.133.

Ganeshmurthy, A.N., Sathisa, G.C. and Prakash Patil. 2011. Potassium nutrition on yield and quality of fruit crops with special emphasis on banana and grapes. Karnataka J. Agric. Sci., 24 (1):(29-38).

Ganry, J. 1975. Effect of covering bunches with polyethylene sleeve on the fruit temperature in the conditions of Guadeloupe. Fruits, 30(12):735-738.

Garcia R., Leal, F. and Rolz, C. 2007. Drying of bananas using microwave and air ovens. International Journal of Food Scicence and Technology, 23(1):73-80.

Garcia, E. and Lajolo, F. M. 1988. The amylase and glucosidase behaviour. Jou. Food Sci. 53:1181-1186.

Garcia, R., Villatoro, L. and Orantes, H. 1987. Reduction of the time required to lower the pH of banana puree by lactic fermentation. Proceedings of Latin American symposium on Biotechnology for the production of biomass and waste treatments. 407-413.

Gaudreau, L., Charbonneau, J., Ve´zina, L.P., Gosselin, A. 1994. Photoperiod and photosynthetic photon flux influence growth and quality of greenhouse grown lettuce. Hortscience 29:1285-1289.

Gayatri G.K. and S.K. Bala. 2006. Non-Traditional natural textile fibres: Production and applications, Asian Textile Journal, 15(11):63-66.

Ghosh, B.K., Patro, B. and Barik, B.C. 1997. Effect of ripening agents and storage conditions on the postharvest changes in banana fruits var. Champa. Orissa Journal of Horticulture 25(1):14-17.

Giami, S.Y. 1991. Effects of pretreatments on the texture and ascorbic acid content of frozen plantain pulp (*Musa paradisiaca.* L.). Journal of the Science of Food and Agriculture. 55(4):661-666.

Ginsburg, L. 1970. What is fruit quality? Deciduous Fruit Grower, 20:248-252.

Giovannoni, J.J. 2004. Gentic regulation of fruit development nad ripening.The Plant Cell, 16:S170-S180.

Goel, R.K. and Sairam, K. 2002. Anti-ulcer drugs from indigenous sources with emphasis on *Musa sapientum*, *Tamrabhasma*, *Asparagus racemosus* and *Zingiber officinale.* Indian J Pharmacol, 34: 100–110.

Goldstein, J.L. and Wick, E.L. 1970. Properties of an enzyme from bananas J Food Sci, 35:482.

Gonza´ lez-Soto, R. A., Mora-Escobedo, R., Hernandez-Sanchez, H., Sanchez- Rivera, M. and Bello Perez, L.A. 2007. The influence of time and storage temperature on resistant starch formation from autoclaved debranched banana starch. Food Research International, 40(2): 304–310.

Gonzalez-Montelongo, R., Gloria Lobo, M., Gonzalez, M. 2010. Antioxidant activity in banana peel extracts: testing extraction conditions and related bioactive compounds. Food Chem 119:1030–9.

Gopalan, C., Rama Sastri, B.V. and Balasubramanian, S.C.1985.Nutritive Value of Indian Foods. NIN, ICMR, Hyderabad.

Gore, M.A. and Akolekar, D. 2003. Evaluation of banana leaf dressing for partial thicknesss burn wounds, Burns, 29:487-92.

Gorris, L.G.M. and Tauscher, B. 1999. Quality and safety aspects of novel minimal processing technologies. In F. A. R. Oliveira and J.C. Oliveira (Eds.), Processing foods. Quality optimization and process assessment (pp. 325–339). New York: CRC Press.

Gottreich, M. and Halevy, Y. 1982. Delaying ripening of pre-harvest bananas (cv.Dwarf Cavendish) with gibberlins. Fruits, 37(2):97-102.

Gowen S 1995. Bananas and plantains. Chapman and Hall, London. 567p.

Gross, K.C., Wang, C.Y., Saltveit, M., 2002. The Commercial Storage of Fruits, Vegetables, and Florist and Nursery Crops. An Adobe Acrobat pdf of a draft version of the forthcoming revision to US Department of Agriculture, Agriculture Handbook 66 on the website of the USDA, Agricultural Research Service, Beltsville Area.

Guenova, E., Hoetzenecker, W., Kisuze, G., Teske, A., Heeg, P., Voykoc, B., Hoetzenecker, K., Schippert, W. and Moehrle, M. 2013. Banana leaves as an alternative wound dressing. Dermatol Surg, 39(2):290-297.

Guilbert, G. and Biquet, B. 1996. Edible Films and Coatings. In G. Bureau & J. L. Multon (Eds.). Food Packaging Technology, Vol. 1:315–353, New York: VCH Publishers Inc.

Guimarães, J.L., Frollini E., da Silva C.G., Wypych F., Satyanarayana K.G. 2009. Characterization of banana, sugarcane bagasse and sponge gourd fibers of Brazil. Industrial Crops and Products 30:407–415.

Guimarães, J.L., Satyanarayana, K.G., Wypych, F. and Ramos, L.P. 2006. Brazilian Patent Provisional No. P10602428-9 dated 28/03/2006 (in Portuguese).

Haard N.F. 1967. The isolation and partial characterization of the mitochondrial fraction from the pulp of ripening banana. Ph.D Thesis, University of Massachusetts.

Haard, N.F. 1973. Upsurge of Particulate Peroxidase in Ripening Banana Fruit. Phytochemistry, 12(3):555-560.

Hailu, M., Workneh, T. S. and Belew, D. 2013. Review on postharvest technology of banana fruit. African Journal of Biotechnology 12(7):635-647.

Halliwell, B. and Gutteridge, J.M.C. 1989. Free Radicals in biology and medicine, Clarendon Press, Oxford, UK, 1989, pp. 51–57.

Hammond, J. Brent, Egg Richard, Diggings Drew and Cable G. Charlie. 1996. Alcohol from bananas., Bioresoune Technology 56: 125-130.

Hao, X. and Papadopoulos, A.P. 1999. Effects of supplemental lighting and cover materials on growth, photosynthesis, biomass partitioning, early yield and quality of greenhouse cucumber. Sci. Hortic. 80:1-18.

Harish, S., Kavino, M., Kumar, N., Saravanakumar, D., Soorianathasundaram, K. and Samiyappan, R. 2008. Applied Soil Ecology, 39:187–200.

Harnly, J.M., Doherty, R.F., Beecher, G.R., Holden, J.M., Haytowitz, D.B., Bhagwat, S. and Gebhardt S. 2006. Flavonoid content of U.S. fruits, vegetables and nuts. J Agric Food Chem 54:9966–77.

HarvestPlus. 2007. Addressing micronutrient deficiencies in Sub-Saharan Africa through Musa-based foods. Annual Technical Report Prepared by Bioversity International With partners: CARBAP – Centre Africain de Recherches sur Bananiers et Plantains, Cameroon KULeuven, Katholieke Universiteit Leuven,Belgium.

Hazarika, D.N. and Mohan, N.K. 1990. Effect of level of nitrogen and time of application on fruit quality of Jahaji banana. Banana Newsletter, 13:26-29.

Hazarika, D.N. and Mohan, N.K. 2007. Growth, yield and quality of banana 'Jahaji' under high density planting. In: Banana-Technological Advancements (Eds. Singh, H.P. and Uma, S). AIPUB, Trichy. Pp.270-272.

Hegde, D. M. 1988. Growth and yield analysis of Robusta banana in relation to soil water potential and nitrogen fertilization. *Sci. Hort.*, 37: 145-155.

Heikal, S.O. and Fadl, M.H. 1977. Mild pulping of banana stem, Research and Industry, 22(4):222.

Heikal, S.O. 1976. Nitric acid paper pulps from banana stem, Indian pulp and paper, 31(3): 5.

Hesselman, C.W. and Frebairn, H.T. 1986. Rate of ripening of initiated bananas as influenced by oxygen and ethylene. J Am Soc Hortic Sci 94:635-637.

Higa, T. 1996. An Earth Saving Revolution. (English translation.) Sunmark Publishers, Inc. Tokyo, Japan.

Hoda Jafarizadeh Malmiri, Azizah Osman, Chin Ping Tan, and Russly Abdul Rahman. 2012. Effects of Edible Surface Coatings (Sodium Carboxymethyl Cellulose, Sodium Caseinate and Glycerol) on Storage Quality of Berangan Banana (*Musa Sapientum* Cv. Berangan) using Response Surface Methodology. Journal of Food Processing and Preservation, 36(3):252-261.

Hofman, P.J. and Smith, L.G. 1993. Preharvest effects on postharvest quality of subtropical and tropical fruit. In: Proceedings of an international conference held at Chang Mai, Thailand, 19-23 July. ACIAR Proceedings. No 50.

Holbery, J. and Houston, D. 2006. Natural-fiber-reinforced polymer composites applications in automotive applications, JOM, 58(11):80–86.

Hong, J.H. and Gross, K.C. 1998. Surface sterilization of whole tomato fruit with sodium hypochlorite influences subsequent postharvest behavior of fresh-cut slices. Postharvest Biology and Technology, 13:51–58.

Hörtensteiner, S. 2006. Chlorophyll degradation during senescence. Ann. Rev. Plant Biol., 57(1):55–77.

Hosam El-Deen, A.S.H., 2002. Effect of sulphur soil application on growth, yieldand fruit quality of hindi banana cultivar. J. Agric. Sci. Mansoura Univ., 27(3):1675-1681.

Hovi-Pekkanen Tiina, and Tahvonen Risto. 2008. Effects of interlighting on yield and external fruit quality in year round cultivated cucumber. Scientia Horticulturae, 116: 152-161.

http://bananasweb.com/bananas/Health+Benefits+of+Bananas accessed on 26.06.2010.

http://doctorsdunia.in/health-benefits-of-banana/ accessed on 26.06.2010.

http://healthmad.com/nutrition/10-health-benefits-of-bananas/ accessed on 26.06.2010.

http://www.technopak.com/Files/Packaged_Juice_Market_in_India.pdf

Huang, X.K. and Chen, C.G. 1993. A study on banana storage. Journal of Fruit Science. 10(4): 221-223.

Huber, D. J. 1983. Polyuronide degradation and hemiellulose modifications in ripening tomato fruit. J. Am. Soc., Hortic. Sci. 108:405–409.

Huddar, A.G., Chandramouli, H.D., Subbanna, V.C. and Jayaprasad, K.V. 1990. Effect of postharvest application of calcium chloride on ripening of banana cv. Robusta. Crop Research. Hissar. 3(1):35-39.

Hulmes, A.C. 1971. Fruit Juices. In: The Biochemistry of Fruits and their products. Academic Press. London. pp.588-589.

Hussein, A. M.; Ibrahim, A.M.F. and Attia, M.M. 1985. New aspects in delaying postharvest ripening of banana. Annals of Agricultural Science, 30(1): 553-568.

Huxsoll, C.C. and Bolin, H.R. 1989. Processing and distribution alternatives for minimally processed fruits and vegetables. Food Technology., 43:124-128.

Irtwange, S.V. 2006. Application of modified atmosphere packaging and related technology in postharvest handling of fresh fruits and vegetables. Agric. Eng. Inter.: the CIGR Ejournal. Makurdi, Nigeria. 4(8):1-13.

Islam Sariful, M. and Kabir Fazlul, A.H.M. 2001. Effect of postharvest treatments with some coating materials on the shelf life and quality of banana. Pakistan Journal of Biological Sciences, 4:1149:1152.

Jackson, T. and Badrie, N. 2002. Quality changes on storage of Caribbean banana (Musa acuminata) wines: effect of pectolase concentration and incubation period: short communication. Journal of Wine Research. 13: 43-56.

Jacob Nicemol and Prema Parukuttyamma. 2008. Novel process for the simultaneous extraction and degumming of banana Fibers under solid-state cultivation, Brazilian Journal of Microbiology, 39:115-121.

Jacob Nicemol, Niladevi K.N., Anisha, G.S. and P. Prema. 2007. Pineapple and Banana Fibres, Asian Textile Journal, 16(1):38-44.

Jacxsens, L., Devlieghere, F. and Debevere, J. 1999. Validation of a systematic approach to design equilibrium modified atmosphere packages for fresh-cut produce. Lebensm. Wiss. Technol. 32:425-432.

Jacxsens, L., Devlieghere, F., Debevere, J. 2002.Temperature dependence of shelf-life as affected by microbial proliferation and sensory quality of equilibrium modified atmosphere packaged fresh produce. Postharvest Biol. Technol. 26:59-73.

Jadesha, G., Hemachandra Haller, Manish, K. Mondhe, Manjunath Hubballi, Prabhakar, K and Prakasham, V. 2012. Role of plant extracts in inducing the systemic acquired resistance in harvested banana against anthracnose disease. Annals of Biological Research, 3(11): 5413-5419.

Jaffery, E.H., Brown, A.F., Kulrilich, A.C., Keek, A.S., Matusheski, N. and Klein, B.P. 2003. Variation in content of bioactive components in broccoli. J Food Compos Anal, 16:323-330.

Jaleel, A.S., Basappa, S.C., Ramesh, A. and Sreekantiah, R.K. (1979) Development studies on enzymatic processing of banana, Indian Food Packer, 33(1):10-12.

Janisiewicz, W.J. and Korsten, L., 2002. Biological control of postharvest diseases of fruits. Annu. Rev. Phytopathol. 40:411-441.

Jauhari O.S., Mishra, R.A. and Tewari, C.B. 1974. Nutrient uptake of banana. Indian J. Agric.Chem. 7:73-79.

Jeffery, P.B. and Banks, N.H., 1994. Firmness-temperature coefficient of kiwifruit. N Z J. Crop Hort. Sci. 22: 97–101.

Jeger, M.J., Eden-Green, S., Thresh, J.M., Johanson, A., Waller, J.M. and Brown, A.E. 1995. Banana diseases. In: Gowen, S.R. (ed.) Bananas and Plantains, pp. 317-381. Chapman and Hall, London.

Jenkins, D.J.A., Wolever, T.M.S., Taylor, R.H., Barker, H., Fielder, H., Baldwin, J.M., Bowling, A.C., Newman, H.C., Jenkins, A.L. and Goff, D.V. 1981. Glycemic index of foods: a physiological basis for carbohydrates, exchange. American Journal of Clinical Nutrition 34:362–366.

Jha Ritu. 2006. Role of calcium in physiology of plants- A Review. Agric. Rev., 27(4):235-246.

Jiang, Y. and Li, Y. 2001. Effects of chitosan on postharvest life and quality of longan fruit. Food Chemistry, 73(2):139-143.

Jiang, Y., Joyce, D.C. and Macnish, A.J. 1999. Extension of the shelf life of banana fruit by 1-methylcyclopropene in combination with polyethylene bags. Postharvest Biology and Technology, 16:187–193.

John, G.G. and Scot, K.J. 1989. Delayed harvesting of bananas with sealed covers on bunches. Effect on fruit yield and quality. Aust.J.Exp.Agric., 29(5):719-726.

Johnston, J.W., Hewett, E.W., Banks, N.H., Harker, F.R. and Hertog, M.L.A.T.M., 2001. Physical change in apple texture with fruit temperature: effects of cultivar and time in storage. Postharvest Biol. Technol. 23:13-21.

Jones, D.R. and Stover, R.H. 2000. Fungal diseases of banana fruit, pre-harvest diseases, *In:* Diseases of Banana, Abaca, D.R.Enset, Jones (Ed.), Wallingford, UK, CABI Publishing, 173-190.

Jones, D.R.1991. Chemical control of crown rot in Queensland bananas. Australian Journal of Experimental Agriculture, 31(5):693-698.

Joshi, S.V., Drzal, L.T., Mohanty, A.K. and Arora, S., 2004. Are natural fiber composites environmentally superior to glass fiber reinforced composites? Composite: Part A 35(3): 371–376.

Juarrez- Garcia, E., Agama-Acevedo, E., Sayayo-Ayerdi, S.G., Rodriguez-Ambriz, S.L. and Bello-Perez, L.A. 2009. Composition, digestibility and application in breadmaking of banana flour. Plant Foods for Human Nutrition, 61(3):131-137.

Kachru, R.P., Balasubramanian, D. and Kotwaliwale, N. 1996. Design, development and evaluation of rotary slicer for raw banana chips. Agricultural Mechanization in Asia, Africa and Latin America. 27(4):61-64.

Kader, A.A. 1992. Postharvest technology of Horticultural crops. University of California, division of Agriculture and Natural Resources, (quality and safety factors, definition and evaluation for fresh horticultural crops) second edition, publication No 3311. pp. 228-345.

Kader, A. 2005. Banana: Recommendations for Maintaining Postharvest Quality. <http://rics.ucdavis.edu/postharvest2/Produce/ProduceFacts/Fruit/banana.shtml>(accessed 13.5.2007).

Kader, A. A. and Watkins, C.B. 2000. Modified Atmosphere Packaging – Toward 2000 and Beyond. HortTechnology, 10:483–486.

Kader, A.A., Morris, L.L., Stevens, M.A. and Albright-Holten, M., 1978. Composition and flavor quality of fresh quality of fresh market tomatoes as influenced by some postharvest handling procedures. J. Am. Soc. Hortic. Sci. 103:6–13.

Kader, Adel. A and Rosa S. Rolle. 2004. The role of post-harvest management in assuring the quality and safety of horticultural produce. Agricultural Servcie Bulletin, FAO, Rome, pp. 207-220.

Kajuna, STAR; Bilanski, W.K. and Mittal, G.S. 1997. Textural changes of banana and plantain pulp during ripening. Journal of the Science of Food and Agriculture. 75(2): 244-250.

Kanazawa, K and H. Sakakibara, 2000. High content of dopamine, a strong antioxidant, in Cavendish banana. J. Agricul. and Food Chem. 48:844-848.

Kane, J.G. and Marathe, G.K. 1949. Preparation of pulp from banana stalk, Journal of the Indian Chemical Society, Industrial News Edition 12:113.

Kanellis Angelos K., Solomos Theophanes, and Mattoo Autar K.1989. Changes in Sugars, Enzymic Activities and AcidPhosphatase lsoenzyme Profiles of Bananas Ripened in Air or Stored in 2.5%02 with and without Ethylene1 Plant Physiol, 90:251-258.

Kashaija, I.N, Speijer, P.R, Gold, C.S and Gowen, S.R. 1994. Occurrence, distribution and abundance of plant parasitic nematodes of bananas in Uganda. African Crop Science Journal 2: 99-104.

Kays, S.J. 1997. Postharvest physiology of perishable plant products. Exon Press Athens, GA. US. 578p.

Khan, Md. Asaduzzaman, Tania Mousumi, Amin S.M. Ruhul, Alam Nadia and NazimuddinMd. 2008. An investigation on the Nutritional Composition of Mushroom (*Pleurotusflorida*) cultivated on different substrates, Bangladesh J. Mushroom, 2(2): 17-23.

Khanna, P. and Garcha, H.S. 1981. Introducing the cultivation of Pleurotus florida in the plains of India. Mush. Sci., 11:691-695.

Kharche, D.S., Kalmegh, V.B., Thakare, S.K. 1996. Development of banana firmness meter. Abstracts of paper of Symposium on Technological advancement in banana/ plantain production and processing. pp.51.

King, A.D., Magnuson, J.A., Torok, T., Goodman, N. 1991. Microbial flora and storage quality of partially processed lettuce. J. Food Sci., 56:459-461.

Koffi, E.K., Sims, C.A. and Bates, R.P. 1991. Viscocity reduction and prevention of browning in the preparation of clarified banana juice. Journal of Food Quality. 14(3): 209-218.

Kohli, R.R., Chacko, B.K. and Randhawa, G.S. 1976. Effect of spacing and nutrition on growth and fruit yield of Robusta banana. Indian J. Agric. Sci. 46: 382-86.

Kohli, R.R. *et al.*, 1984. Indian J. Hort., 41:194-198

Kolekar, T.G., Modak, H.M. and Jadhav, S.J. 1988. Shelf-life extension of banana by use of sucrose ester formation. Indian Journal of Plant Physiology, 31:16–20.

Korsten, L., De Villiers, E.E., Wehner, F.C., Kotze, J.M., 1997. Field sprays of *Bacillus subtilis* and fungicides for control of preharvest fruit diseases of avocado in South Africa. Plant Dis. 81:455-459.

Korsten, L., De Villiers, E.E.,Wehner, F.C., Kotze, J.M., 1997. Field sprays of *Bacillus subtilis* and fungicides for control of preharvest fruit diseases of avocado in South Africa. Plant Dis. 81:455–459.

Krauss, U and Johanson, A. 2000. Recent advances in the control of crown rot of banana in the Windward Islands. Crop Protection 19:151-160.

Krauss, U., Operation 2000/Geest, 1996a. Improved fruit quality by clustering in the shed. Farm Management Information No. 1/96, WIBDECO Technical Services Division, St. Lucia.

Krauss, U. and St. Rose, M. 1996. Improved crown rot control in the field and in the shed. Fact Sheet No. 1, WIBDECO Technical Services Division, St. Lucia.

Krokida, M. K. and Maroulis, Z. B. 1999. Effect of microwave drying on some quality properties of dehydrated products. Drying Technology, 17(3): 449-466.

Krokida, M. K., Karathanos V. T. and Maroulis Z. B. 1998a. effect of freeze-drying conditions on shrinkage and porosity of dehydrated agricultural products. J. Food Eng. 35(4): 369-380.

Krokida, M. K., Tsami, E. and Maroulis, Z.B. 1998b. Kinetics of color change during drying of some fruits and vegetables. Drying Technology, 16(3-5): 667-685.

Kulkarni, A.G., Satyanarayana, K. G., Rohatgi, P. K. and Vijayan, K. 1982. Mechanical Properties of Banana Fibers, Journal of Material Science, 18:2290-2296.

Kumah, P., Ampomah, E.O., Olympio, N.S. and Moses, E. 2010. Efficacy of four botanicals and two chemical fungicides in control of crown rot disease of banana (Musa sp., AAA) 'Medium Cavendish. Acta Horticulturae, (II All Africa Horticulture Congress, ISHS).

Kumar Ajay and Mishra Sanjay. 2010. Studies on Production of Alcoholic Beverages from some Tropical Fruits. Indian Journal of Microbiology, 50(1):88-92.

Kumar Dinesh, Pandey, V. and Anjaneyulu, K. 2008. Effect of planting density and nutrient management on growth, yield and quality of micro-propagated banana Rasthali Pathkapoora (AAB). Indian Journal of Horticulture, 65(3):272- 276.

Kumar Manish and Kumar Deepak. 2011. Comparative study of pulping of banana stem. International Journal of Fiber and Textile Research, 1(1):1-5.

Kumar, R., Choudary, V., Mishra, S. and Varma, I.K. 2008. Banana Fiber Reinforced Biodegradable Soy Protein Composites, Frontiers of Chemistry in China, 3(3): 243-250.

Kumar, Sanjay and Sinha Suresh K. 1992. Alternative Respiration and Heat Production in Ripening Banana Fruits (*Musa paradisiaca* van *Mysore Kadali*). Journal of Experimental Botany, 43(12): 1639-1642.

Lahav, E. 1972. Effect of different amount of potassium on growth of banana. Trop. Agrl., (Trinidad), 49: 321-335.

Lampe Johanna W. 1999. Health effects of vegetables and fruit: assessing mechanisms of action in human experimental studies1–3.1999. Am J Clin Nutr 1999;70 (suppl):475S–90S.

Langkilde, A. M., Champ, M. and Andersson, H. 2002. Effects of high-resistant-starch banana flour (RS2) on in vitro fermentation and the small-bowel excretion of energy, nutrients, and sterols: an ileostomy study. American Journal of Clinical Nutrition, 75:104–111.

Le Dividich, J., Geoffroy, F., Canope, I. and Chensot, M. 1978. FAO Animal Production and Health Paper: Using waste bananas as animal feed.Ruminant nutrition: selected articles from the World Animal Reviews. pp160.

Lee, K.Seung and Adel A. Kader. 2000. Preharvest and postharvest factors influencing vitamin c content of horticultural crops. Postharvest Biology and Technology, 20: 207-220.

Lee, L., Arul, J., Lencki, R., Castaigne, F. 1995. A review on modified atmosphere packaging and preservation of fresh fruit and vegetables: physiological basis and partical aspects – Part 1. Packaging Technology and Science. 8:315-331.

Leon, Raul. 1997. Aplicación del Sistema TDF para la remediación bioenzimática de las lagunas facultativas de la Compañia Munidimar, S.A. Técnica del Futuro, S.A. San José, Costa Rica.

Leslie, C.A. and R.J. Romani. 1986. Salicylic acid: a new inhibitor of ethylene biosynthesis, Plant Cell Rep. 5:144–146.

Leslie, C.A. and Romani, R.J. 1988. Inhibition of ethylene biosynthesis by salicylic acid, Plant Physiol. 88:833–837.

Levi A., Ramirez-Martinez, J.R. and Padua H. 1980. Influence of heat and sulphur dioxide treatments on some quality characteristics of intermediate-moisture banana. International Journal of Food Science and Technology, 15(5):557-566.

Linko, M., Haikara, A., Ritala, A. and Penttilä, M. 1998. Journal of Biotechnology, 65:85–98.

Llyas, M.B., Ghazanfar, M.U., Khan, M.A., Khan, C.A and Bhatti, M.A.R. 2007. Post-harvest losses in apple and banana during transport and storage, Pak.J.Agri.Sci, 44:534-539.

Lockard, R.G 1975. The effect of growth inhibitors and promoter on the growth, flowering and fruit size of banana plants. Malaysian Agric. Res., 4(1):19-29.

Lopez, A.1998. Bananas and plantain. CORBANA, Corporacion Bananera Nacional. San Jose, C. R.

Lun, Z.R., Burri, C., Menzinger, M. and Kaminsky, R. 1994. Antiparasitic activity of diallyl trisulfide (Dasuansu) on human and animal pathogenic protozoa (Trypanosoma sp. Entamoeba histolytica and Giardia lamblia) in vitro. Annales de la Socie´te´ Belge de Me´decine Tropicale, 74:51–59.

Luntala, T., Mbile, F., Jean Claude, O., and Okulo, M. 2000. *Faculty of Agriculture and Applied Biological Sciences ICP/food science and nutrition*: 1999-2004; Country Presentation. DRC.

Luo, Y. 2005. Emerging technologies for improving food safety and quality of fresh-cut produce—Acidified sodium chlorite and 1-methylcyclopropeneInstitute of Food Technologists meeting abstracts 2005. 17 May 2006.

Lurie, S., 1998. Postharvest heat treatment. Postharvest Biol. Technol. 14:257–269.

Lyons, J.M. 1973. Chilling injury in plants. Ann. Rev.Plant Physiol., 24:445-466.

Madhava Rao, D. and Rama Rao, M. 1979. Post harvest changes in banana cv.Robusta V.N. (1974). Indian J. Hort. 36 (4):387-393.

Mahalakshmi, M. and S. Sathiyamoorthy, 1999. Effect of post shoot application of potassium on yield characters in banana cv. Rasthali (AAB). South Indian Horticulture, 47(1/6): 155-157.

Mahmoud, G.A., Kareem, M.K. El-Tobgy and M.A. Abo-El-Seoud. 2010. Biocide application for increasing storage periods and maintaining quality and chemical constituents of banana fruits. Archives of Phytopathology and Plant Protection, 43(9): 910-921.

Mahouachi, J. 2007. Growth and mineral nutrient content of developing fruit on banana plants (*Musa acuminata* AAA, 'Grand Nain') subjected to water stress and recovery. J. Hort. Sci. Biotech. 82: 839-844.

Maini, S.B. 1999. Horticultural Waste-Ideas for Value Addition. Agriculture and Industry Survey, August. 32.

Manach C, Scalbert A, Morand C. 2004. Polyphenols: food sources and bioavailability. Am J Clin Nutr 79:727–47.

Mandal, B.K. and S.B. Sharma, 1999. Growth and yield responses of Robusta banana *(Musa* AAA) at high densities. Prog. Hort., 31(3-4): 138-143.

Manimegalai, G., Bairavi, K., Jothilakshmi, K. and Subbalakshmi S.R. 1999. Delicious banana product. Project Report Department of Home Science. Home Science College and Research Institute. Madurai.

Mao, L., J. Jeong, F. Que, and D.J. Huber. 2006. Physiological properties of fresh-cut watermelon (Citrullus lanatus) in response to 1- methylcyclopropene and post-processing calcium application. J. Sci. Food Agr. 86:46–53.

Mapson, L.W. and Robinson, J.E. 1966. Relation between oxygen tension, biosynthesis of ethylene, respiration and ripening changes in banana fruit. J Food Technol 1: 215-225.

Maqbool Mehdi., Asgar Ali and Peter G. Alderson. 2010. A combination of gum arabic and chitosan can control Anthracnose caused by Colletotrichum musae and enhance shelf life of banana fruit. Journal of Horticultural Science & Biotechnology, 85(5): 432-436.

Maqbool, Mehdi., Asgar Ali., Senthil Ramachandran, Daniel R. Smith and Peter G. Alderson. 2010. Control of post harvest anthracnose of banana using new edible composite coating, Crop Protection, 29:1136-1141.

Mari, M., Guizzardi, M., 1998. The postharvest phase: emerging technologies for the control of fungal diseases. Phytoparasitica, 26:59–66.

Marin-Rodriguez, M.C., Smith, D.L., Manning, K., Orchard, L. and Seymour, G.B. 2003. Pectate lyase gene expression and enzyme activity in ripening bananas. Plant Mol Biol, 51, 851.

Markovic, D., Heinrichova, K. and Lenkey, B. 1975. *In:* Pectolytic enzymes from banana. Tropical and Subtropical Fruits (S Nagy & PE Shaw, Editors) AVI Pub.Inc. Westport, Connecticut.

Marriott, J. and Lancaster, P.A. 1983. Bananas and Plantains. In: Handbook of Tropical Foods. Harvey Jr. TC (Ed), Marcel Dekker, Inc. pp. 85142.

Marriott, J. 1980. Bananas-physiology and biochemistry of storage and ripening for optimum quality. CRC Critical Reviews of Food Science and Nutrition, 13(1):41-88.

Martinez-Ovanto, M., Sayago-Ayerdi, S.S., Agama-Acevedo, E., Goni, S. and Perez-Bello, L. 2009. Unripe banana flour as an ingredient to increase the undigestible carbohydrates of pasta. Food Chemistry, 113:121-126.

Martin-Prevel, P. 1973. Influence de la nutrition potassique sur les functions physiologiques et al qualite de al production chez quelques plantes tropicales. In 10th Colloq of the International Potash Institute, pp-233-248.

McFeeters, R.F., Barrangou, L.M., Barish, A.O. and Morrison, S.S. 2004. Rapid softening of acidified peppers: effect of oxygen and sulfite. Journal of Agriculture Food Chemistry, 52:4554–4557.

McHugh, T.H. and Senesi, E. 2000 Apple wraps: a novel method to improve the quality and extend the shelf life of fresh-cut apples. J. Food Sci. 65(3): 480-5.

McMurchie, E.J., McGlasson, W.B. and Eaks, I.L. 1972. Treatment of fruit with propylene gives information about biogenesis of ethylene. Nature, 237:235–236.

Mendoza, E.M.T., Laurena, A.G and Uritani, I. 1994. Postharvest Biochemistry of Plant Food-Materials *In:* Postharvest Biochemistry of Plant Food-Materials in the Tropics (I Uritani, V V Grscia & EMT Mendoza, Editors) Japan Scientific Society Press, Tokyo, pp 177-191.

Mengel, K. and Kirkby, E.A. 1987. Principles of Plant Nutrition. 4th Edition. International Potash Institute, IPI, Bern, Switerzerland, pp.685.

Miller V. Erston and Heilman, S. Alan. 1952. Ascorbic Acid and Physiological Breakdown in the Fruits of the Pineapple (Ananas comosus L. Merr.) Science, 116(3019): 505-506.

Mishra, S.D., Desai, B.M. and Gaur, B.K.1981. Effect of gibberlic acid spraying on banana fruit development. Current Science, 50: 275-277.

Mohapatra Debabandya, Sabyasachi Mishra and Namrata Sutar. 2010. Banana Post Harvest Practices: Current Status and Future Prospects- A Review. Agric. Rev., 31(1):56-62.

Mokbel Matook Saif and Fumio Hashinaga, 2005. Antibacterial and Antioxidant Activities of Banana (*Musa*, AAA cv. Cavendish) Fruits Peel. American Journal of Biochemistry and Biotechnology 1(3):125-131.

Molla, M.M., Nasrin, T.A.A. and Nazrul-Islam, M. 2009. Study on the suitablility of banana varieties in relation to preparation of chips. J. Agric. Rural Dev. 7:81-86.

Morton, J.F. 1987. Banana. *In*: Fruits of warm climates. Miami, Florida. pp. 29-46.

Mostafa, E.A.M. and Abd El-Kader. 2006. Sulfur Fertilization Effects on Growth, Yield and Fruit Quality of Grand Nain Banana Cultivar. *J. Applied Sci. Res.* 2:470-476.

Mostafa, E.A.M. 2005. Response of Williams banana to different rates of nitrogen and potassium fertilizers. Journal of Applied Sciences Research, 1: 67-71.

Mota, R. V., Lajolo, F. M., Ciacco, C. and Cordenunsi, B.R. 2000. Composition and functional properties of banana flour from different varieties. Starch/Sta¨rke, 52(2-3):63-68.

Mukherjee, K.G. and Satyanarayana, K.G. 1984. Structure and Properties of some Vegetable Fibers, Journal of Material Sciences, 19: 3925-3934.

Mukhopadhyay, S., Fanguerio, R., Arpac, Y. and Sentrik, U. 2008. Banana fibres-variability and fracture behaviour, J.Engineered Fibres and Fabrics, 3(2):39-45

Munasque, V.S., Abdullah, H., Gelido, M.E.R.A., Rohaya, M.A. and Zaipun, M.Z. 1990. Fruit growth and maturation of banana, pp.33-43. In: Hassan, A. and E.B. Pantastico, 1990. Banana fruit development, postharvest physiology, handling and marketing in ASEAN.

Murray, D.B.1959. Deficiency symptoms of the major elements in the banana. Trop.Agric., Trinidad, 36:100-107.

Mustaffa, M.M. 1983. Effect of spacing, and nitrogen on growth, yield and quality of hill banana. South Indian Horticulture, 31:270-273.

Mustaffa, M.M. 1988. Effect of spacing and nitrogen on growth, fruit and yield of Robusta banana grown under rainfed conditions. South Indian Hort., 36(5):228-231.

Mustaffa, M.M., Tanuja Priya, B., Sivakumar, K.C., Kumar, V. and Sathiamoorthy, S. 2007. Effect of pre-harvest treatments on bunch parameters, quality and shelf life of banana Nendran. In : Banana-Technological Advancements (Eds.H.P.Singh and S.Uma), AIPUB, NRC for Banana, Trichy, pp.474-478.

Muyonga, J.H. 2000. Production and evaluation of precooked dehydrated unripe banana slices. Afr.Crop Sci., J. 8:1022-1027.

Nabeesa, B. and Unikrishnan, S.C. 1988. Carbohydrates, amylase and invertase activity during ripening in Red and Green Red banana. Indian Journal of Plant Physiology. 31(1):28-37.

Nachene, R.P. and Uma, S. 2003. Processing of Banana pseudostem for extraction of fibres. Book of Abstract of 6th Agricultural Science Congress. February 13-15th 2003, Bhopal. India. pp 254.

Nagaraju, C.G., Reddy, T.V. 1995. Deferral of banana fruit ripening by cool chamber storage. Advances in Horticultural Science. 9(4): 162-166.

Nalina, L., N. Kumar, S. Sathiamoorthy and Muthuvel, P. 2000. Effect of nutrient levels on bunch characters and vegetative characters of banana cv. Robusta under high density planting system. South Indian Hort.,48(1-6): 18-22.

Namaganda, J.M., Karamura, E., Namanya P., Gowen, S.R. and Bridge, J. 2000. Studies on host range of the banana lesion nematode, *Pratylenchus goodeyi*. In: Craenen, K., Ortiz, R., Karamura, E.B., & Vuylsteke (Eds.) Proceedings of the International Conference on Banana and Plantain for Africa. Acta Horticurae No. 540. pp 419-425. International Society for Horticultural Science (ISHS), Leuven, Belgium.

Nanda, S.K., Vishwakarma, R.K., Bathla, H.V.L., Anil Rai and P. Chandra. 2012. Estimation of quantitative harvest and post-harvest losses of major agricultural produce in India. All India Coordinated Project on Post Harves Technology, CIPHET (ICAR), Ludhiana.

Nandhini, P., Ravi, I. and Narayana, C.K. 2010. Carbohydrate hydrolyzing and antioxidative enzymes in banana fruit. Authors: *In:* Banana New Innovations. (Ed. H.P. Singh and M.M. Mustaffa) Westville Publishing House, New Delhi.

Narayana, C.K. 2003. Annual Report of AP Cess fund project on Pre and Post Harvest Technologies for Export Quality Rasthali and Neypoovan Bananas. National Research Centre for Banana, Trichy.

Narayana, C.K. and Mustaffa, M.M. 2000. In Annual Report. National Research Centre for Banana. Trichy. Tamil Nadu.

Narayana, C.K. and Mustaffa, M.M. 2006. Improved Post Harvest Handling Technology in Banana. Tech. Bull No.15. National Research Centre for Banana, Trichy, Tamil Nadu.

Narayana, C.K. and Mustaffa, M.M. 2007. Influence of maturity on shelf life and quality changes in banana during storage under ambient conditions. Indian J. Hortic., 64(1):12-16.

Narayana, C.K. and Singh, B.P. 1998. Effect of ameliorative treatments on reduction of chilling injury in guava fruits. Prog. Hort., 30(3-4):148-152.

Narayana, C.K., Mustaffa, M.M. and Sathiamoorthy, S. 2002e. Effect of packaging and storage on shelf life and quality of banana cv.Karpuravalli. Indian J.Hort, 59(2): 113-117.

Narayana, C.K., Mustaffa, M.M. and Sathiamoorthy, S. 2002a. Effect of postharvest application of calcium chloride on ripening, shelf life and quality of banana cv. Poovan. South Indian Horticulture, 50(4-6): 308-316.

Narayana, C.K., Mustaffa, M.M. and Sathiamoorthy, S. 2002b. Effect of low temperatures on chilling injury and quality changes in Pachanadan (Pome, AA group) banana. Paper presented in 'Global Conference on Banana and Plantain' held at Bangalore on 29-31st October 2002. Abstract No. P.P.09.5., pp.208.

Narayana, C.K., Sathiamoorhty, S., Mustaffa, M.M. and Evelin Mary, A. 2002g Processing and Value Addition in Banana. National Workshop on Post Harvest Management of Horticultural Produce., organized on 7-8th February 2002 at Hotel Shivalikview, Chandigarh by National Productivity Council.

Narayana, C.K., Sathiamoorthy, S. and Evelin Mary, A. 2002c. Studies on ready-to-serve beverage from enzyme clarified banana juice. Progressive Horticulture, 34(1):65-71.

Narayana, C.K., Sathiamoorthy, S. and Krishnan, P. 2007. Effect of bunch covering on yield and yield parameters of banana (Cv. Neypoovan). *In:* Banana-Technological Advancements (Eds. Singh, H.P. and Uma, S). AIPUB, Trichy. pp. 315-317.

Narayana, C.K., Sathiamoorthy, S. and Ramajayam, D. 2002f. Preliminary studies on preparation of Banana sauce. Paper presented in 'Global Conference on Banana and Plantain' held at Bangalore on 29-31st October 2002. Abstract No. P.P.09.4., pp. 207.

Narayana, C.K., Shivashankar, S., Mustaffa, M.M. and Sathiamoorhty,S. 2002d. Studies on suitability of varieties of banana and frying medium for production of chips. Beverage and Food World. 29(6):29-30.

Narayana, C.K., Sree Prabha, S and Sathiamoorthy, S. 2007. Production of ethanol from waste banana peel. Mysore J. Agric. Sci, 41(2): 282-285.

Narayana, C.K., Sathiamoorthy, S.. Evelin Mary, A. and Ramajayam, D. 2003a. Studies on quality changes of ready-to-serve banana juice during storage. Book of Abstract of 6th Agricultural Science Congress. February 13-15th 2003, Bhopal. India. Pp256.

Narayana, C.K., Ramajayam, D., Evelin Mary, A. and Sathiamoorthy. 2003b. Value Added Banana Products. National Research Centre for Banana, Tiruchirapalli, Tamil Nadu, India.

Narayana, C.K. Sathiamoorthy, S., Evelin Mary, A. and Ramajayam, D. 2002e. Banana Wine – A Novel Alcoholic Beverage. Agro India, Sept- October, 2002, pp.22-23.

Narayana, Cherukatu Kalathil and Micheal Pillay. 2011. Postharvest Processed Products from Banana. In: Banana Breedin - Progress and Challenges (Eds. Michael Pillay and Abdou Tenkouano), CRC Press, Boca Raton, FL 33487-2742, pp269-284.

Narayana. C.K and Priyalakshmi, G. 2008. Banana-Nature's Marvel: Medicinal and therapeutic values –AIPUB, Trichy, 2008.

NCR, 1978. Report of the Steering Committee for study on Postharvest Food Losses in Developing Countries, National Research Council, National Academy of Sciences, Washington, DC.

Ndungo, V., Fiaboe, K.K.M. and Mwangi, M. 2008. Banana Xanthomonas wilt in the DR Congo: Impact, spread and management. Journal of Applied Biosciences, 1(1): 1-7.

Neelakantan, S. and Sadasivam, R. 1977. Ripening behaviour of banana stored in moist sand. South Indian Horticulture. 25(4): 160-161.

Netravali, A.N., Chabba, S. 2003. Composites get greener. Mater. Today 6(4):22–29.

Ngalani, J.A., Signoret, A. and Crouzet, J. 1993. Partial purification and properties of plantain polyphenol oxidase. Food Chem., 48: 341-347.

NHB, 2011. Horticulture Database – 2011. National Horticulture Board, Ministry of Agriculture and Cooperation, Government of India.

NHB, 2012. Horticulture Database – 2012. National Horticulture Board, Ministry of Agriculture and Cooperation, Government of India.

Nishiyama, K., Guis, M., Rose, J.K.C., Kubo, Y., Bennett, K.A.B., Lu, W., Kato, K., Koichiro, U., Ryohei, N., Akitsugu, I., Mondher, B., Alain, L., Jean-Claude, P. and Bennett, A.B. 2007. Ethylene regulation of fruit softening and cell wall disassembly in charentais melon. J. Exp. Bot. 58:1281-1290.

Nisperos-Carriedo, M.O., Baldwin, E.A. and Shaw, P.E. 1992. Development of an edible coating for extending postharvest life of selected fruits and vegetables. Proceedings of Florida State Horticultural Sociey 104:122-125.

Nogueira, R.I. and Park, K.J. 1992. Drying parameters to obtain 'Banana Passa'. International drying Symposium Abstracts. 874-883.

Northover, J. and Schneider, K.E. 1996. Physical mode of acion of petroleum and plant oils on powdery and downy mildews of grapevines. Plant Disease, 80(5): 544-550.

Occena-Po G.Lillian. 2006. Banana, Mango and Passion Fruit. In: Handbook of fruit and frui tprocessing (Eds. Hui, Y.H., Barta, J, Pilar Cano, M, Gusek, T, Sidhu, J.S and Sinha, N.), Blackwell Publishing Professional, Iowa, USA.Pp.635-638.

Odibo, F.J.C., Nwankwo, L.N. and Agu, R.C. 2002. Process Biochemistry, 37, 851–855.

Ogazi, P.O. 1996. Plantain: Production, Processing and Utilization. Paman and Associates Limited, Uku-Okigwe.

Olivas, G. I. and Barbosa-Canovas G. V. 2005. Edible coatings for fresh-cut fruits. Critical Reviews in Food Science and Nutrition 45(8):657-670.

Oluwafemi Ademiluyi Benson. 2013. Influence of number of sucker per plant on the growth, yield and yield components of Plantain (Musa sp.) in Ado-Ekiti, Nigeria. Agricultural Science Research Journals, 3(2):45-49.

Ortega, M.M. and Blanco, C. 1986. Vinegar from banana wastes in Panama. Informe Mensual UPSEB. 10(75): 47-52.

Ortiz, R., Ferris, R.S.B. and Vuylsteke, D.R. 1995. Banana and plantain breeding. pp. 110-145. *In:* Gowen S. (ed). Bananas and plantains, Natural Resources Institute and Department of Agriculture University of Reading, UK, Chapman & Hall.

Osman Ahmed Ismail and Wondimu Asrat. 1985. Sun drying of banana. Ethiopian Nutrition Institute. pp1-18.

Padam B.S., Tin H.S., Chye F.Y. and Abdullah M.I. 2012. Antibacterial and Antioxidative Activities of the Various Solvent Extracts of Banana (*Musa paradisiaca* cv. Mysore) Inflorescences. Journal of Biological Sciences, 12: 62-73.

Palmer, J.K. 1971. The Banana. *In:*The Biochemistry of Fruits and their Products, A.C. Hulme (Ed.), Vol.2, Academic Press, London,65-101.

Pandey, V., Kumar, D and George, S. 2005. Response of micropropagated 'Robusta' banana to varying combinations of N, P, K nutrition in lateritic soils of coastal Orissa. Indian J. Hort. 62(2): 122-126.

Parida, G. N., Ray, D.P., Nath, N. and Dora, D.K. 1994. Effect of graded levels of NPK on growth of Robusta banana. Indian Agric., 38 (1): 43-50.

Parmar, B.R. and Chundawat, B.S. 1984. Effect of growth regulators and sleeving on maturity and quality of banana cv. Basrai. South Indian Hort., 32(4):201-204.

Pascual-Teresa de S., Santos-Buelga C., Rivas-Gonzalo J.C. 2000. Quantitative analysis of flavan-3-ols in Spanish foodstuff and beverages. J. Agric. Food Chem. 48:5331–5337.

Pathak, N., Asif, M.H., Dhawan, P., Srivastava, M.K., Nath, P. 2003.Expression and activities of ethylene biosynthesis enzymes during ripening of banana fruits and effect of 1-MCP treatment. Plant Growth Regulation 40:11–19.

Pathak, N. and Sanwal, G.G. 1998. Multiple forms of polygalacturonase from banan fruits. Phytochemistry, 48(2):249-255.

Paunescu, C. and Bratucu, Gh. 2010. Research regarding the storage of fruits and vegetables in controlled atmosphere. Bulletin of the *Transilvania* University of Brasov., 3(52):177-182.

Pawar, R.D. 1999. Export trial of Banana to Dubai. Souvenir of Seminar on Technological Advancement in Banana Production, Handling and Processing Management, held at Jalgaon from 27-28th March 1999.

Payasi A, Misra, P.C. and Sanwal, G.G. 2004. Effect of phytohormones on pectate lyase activity in ripening *Musa acuminata*. 42:861–865.

Payasi, A and Sanwal,G.G. 2003. Pectate lyase activity during ripening of banana fruit. Phytochem, 63:243-248.

Peacock, B.C.1980. Banana ripening-Effect of temperature on fruit quality. Qld.J.Agric.Anim.Sci., 37(1):39-45.

Pelayo, C., Vilas-Boas, E.V., Benichou, M. and Kader, A.A. 2003. Variability in responses of partially ripe bananas to1-methylcyclopropene. Posth. Biol. Technol., 28:75-85.

Pellegrini, N., Serafini, M., Colombi, B., Del Rio, D., Salvatore, S., Bianchi, M. and Brighenti, F. 2003. Total antioxidant capacity of plant foods, beverages and oils consumed in Italy assessed by three different *in vitro* assay. J Nutr. 133:2812–2819.

Pesis, E. 2004. Respiration and ethylene. International Research and Development on Postharvest Biology and Technology. The Volcani Center, Israel.

Pesis, E., Ben-Arie, R., Feygenberg, O. and Villamizar, F. 2005. Ripening of ethylene-pretreated bananas is retarded using modified atmosphere and vacuum packaging. Postharvest Biology and Technology, 40:726–731.

Pesis, E., Copel, A., Ben-Arie, R., Feygenberg, O. and Aharoni, Y. 2001. Low oxygen treatment for inhibition of decay and ripening in organic banana. The Journal of Horticultural Science and Biotechnology, 76:648–652.

Pheantaveerat, A. and Anprung, P. 1993. Effect of pectinases, cellulases and amylases on production of banana juice. Food 23(3):188-196.

Philippe, Tixier., Bugaud Christophe., Duguet Remy and Salmon Frederic. 2010. Effect of pre-harvest and post harvest application of calcium on banana green life. Fruits., 65(4): 201-208.

Pokharkar, S.M. and Prasad, S. Mass transfer during osmotic dehydration of banana slices. Journal of Food Science and Technology, 35(4):336-338.

Poonguzhali, P.K. and Chegu, H. 2004. Department of Medical Biochemistry, Dr. A L M Postgraduate Institute of Basic Medical Sciences, University of Madras, Taramani, India.

Pothavorn, P., Kitdamrongsont, K., Swangpol S., Wongniam, S., Atawongsa. K., Savasti, J. and Somana, J. 2010. Sap phytochemical compositions of some bananas in Thailand. J Agric Food Chem., 11; 58(15):8782-7.

Pradhan, S.K., Bandyopadyay, A., Mitra, S.K. and Sen, S.K. 1988. Effect of growth substances on fruit size, yield and quality of banana cv.Gaint Governor., Prog. Hort., 20(3-4):326-330.

Prasad K.V., Bharathi K. and Srinivasan, K.K.1993. Evaluation of Musa (Paradisiaca Linn. cultivar)—"Puttubale" stem juice for antilithiatic activity in albino rats. Indian J Physiol Pharmacol. 1993 Oct; 37(4):337-41.

Promyou, S., Ketsa, S. and Van Doorn, W.G. 2007. Effect of surface coating on ripening and early peel spotting in 'Sucrier' banana (*Musa acuminata*), New Zealand Journal of Crop and Horticultural Science, 35(2):259-265.

Pua E.C., Ong C.K., Liu P., Liu J.Z., 2001.Isolation and expression of two pectate lyase genes during fruit ripening of banana. (*Musa acuminata*). Plant, 113:92-99.

Purvis, A.C. 1983. Effects of film thickness and storage temperature on water loss and internal quality of seal-packaged grapefruit. Journal of American Society of Horticultural Science, 108:562–566.

Quazi, M.H., Freebairn, H.T. 1970. The influence of ethylene, oxygen and carbon dioxide on the ripening of bananas. Bot Gaz 131:5-14.

Raghupathi, H.B., Srinivas, K. and Reddy, B.M.C. 2002. Concentration and distribution of secondary and micro-nutrients in banana under fertigation. South Indian Hort., 50(4-6): 291-300.

Ragsdale, N.N., Sisler, H.D., 1994. Social and political implications of managing plant diseases with decreased availability of fungicides in the United States. Annu. Rev. Phytopathol. 32:545–557.

Ram, R. A. and Prasad, J.1988. Effect of nitrogen, phosphorus and potassium on fruit quality of banana cv. Campirganj Local (Musa-ABB). South Indian Hort. 36: 290-292.

Ram, R.A.J. and Prasad, J. 1988. South Indian Hort., 36:290-292

Ramanathan, G., Vadivelu, S., Krishnamoorthy, K.K., Helkiah, J and Shetty, K.S. 1973. Effect of ammonium chloride and ammonium sulphate on banana. Madras Agric. J., 60:1066-1067.

Ramesh Kumar, A., Kumar, N. and Kavino, M. 2006. Role of potassium in fruit crops- A Review. Agric. Rev., 27(4):284-291.

Ramirez-Martinez, J.R., Levi, A., Padua, H. and Bakat, A.1977. Astringency in an intermediate moisture banana product. J.Food Sci., 42:1201-1203.

Rao, D.V.R. and Chundawat, B.S. 1991. Chemical regulation of ripening in banana cv. 'Lacatan' at non-refrigerated temperatures. Haryana Journal of Horticultural Sciences. 20(12): 6-11.

Raskin, A. Ehmann, W.R. Melander, B.J.D. Meeuse. 1987. Salicylic acid: a natural inducer of heat production in *Arum lillies*, Science 237:1601-1602.

Raskin, I. 1992a. Role of salicylic acid in plants, Ann. Rev.Plant Physiol. Mol. Biol. 43:439–463.

Raskin, I.1992b. Salicylate, a new plant hormone, Plant Physiol. 99:799–803.

Ray, D.P., Bhadurim, S.K., Nayak, L.K., Ammayappan, L., Manna, K and Das, K. 2012. Utilization and value addition of banana fibre- A Review. Agricultural Reviews., 33(1):46-53.

Reddy, B.M.C. and Patil, Prakash. 2007. High density planting in banana. In: Banana-Technological Advancements (Eds.) Singh, H.P. and Uma, S. AIPUB, Trichy. pp. 273-283.

Risse, L.A. and Miller, W.R. 1983. Film wrapping and decay of eggplant. Proceedings of Florida State Horticultural Society 96:350-352.

Rivera, A., González, J.S., Carrillo, R. and Martínez, J. M. 2009. Journal of Cleaner Production, 17(2):137-142..

Robinson, J.C. 1996. Bananas and plantains. Wallingford, UK: CABI Publishing, CAB International.

Robinson, A. A. 1980. Research Design and development of banana dehydration process. Food Engineering,. Sydney, UNSW, Australia.

Robinson, J.C. and Human N.B. 1988. Forecasting of banana harvest ('Williams') in the subtropics using seasonal variation I bunch development rate and bunch mass. Scientia Horticulturae 34(3-4):249-263.

Robinson, J.C. and Nel, D.J. 1984. Influence of polyethylene bunch covers on yield and fruit quality of winter –developing banana bunches. Hortic.Science (South Africa), 1:26-28.

Rodriquez, P.V., Guerra, D., Manzano, J. and Campbell, R.J. 1995. Precooling of banana fruits (Musa ABB banana subgroup) under different storage conditions. Proceedings o the Interamerican Society for Tropical Horticulture, 39:94-99.

Rolle, R.S. and Chism, G.W. 1987. Physiological consequences of minimally processed fruits and vegetables. J. Food Qual. 10:157-177.

Romero -José Orozco and Zamora -Octavio Pérez. 2006. Soil Moisture Tension and Nitrogen Fertilization on Banana (Musa AAA Simmonds) Cv. Gran Enano. Agrociencia 40(2):149-162.

Rose JKC, Lee HH and Bennett AB 1997. Expression of a divergent expansin gene is fruit-specific and ripeningregulated. Proc. Natl. Acad. Sci. USA 94: 5955-5960.

Rouhani, I. 1978. Effect of silver ion application on the ripening and physiology of stored green bananas. Journal of Horticultural Science. 53(4): 317-321.

Ryle, G. Respiration and plant growth. In: The Physiology and Biochemistry of Plant Respiration. (Ed. Palmer, J). Cambridge University Press, Cambridge. 1984.

Sadasivam, S. and Muthuswamy, S. 1973. Regulate banana ripening. Indian Horticulture. 18(1): 20.

Sadhu, B.P., De, A.B. and Gupta, K.1992. A new post-harvest disease of banana. Indian Phytopathology, 45:473.

Saikia, D.C. et al, 1997. Wild banana plant as source of fiber for paper and cordage industries , J. Sci. Ind. Res. 56(7):408.

Salunkhe DK, Kadam SS 1995. Banana. pp. 67-90. Hand Book of Fruit Science and Technology. Marcel Dekker, Inc. New York.

Sanchez Nieva, F., Hernandez, I. And Bueso de Vinas, C. 1971. Studies on the ripening of plantains under controlled conditions. J.Agric. University. Puerto Rico, 54(3):517-529.

Sangeetha Ganesan, Raman Thangavelu and Swaminathan Usharani. 2010. Evaluation of plant oils for suppression of Crown rot disease and improvement of Shelf life of banana (*Musa* spp. AAA Subgroup cv. Robusta). International Journal of Food Science and Technology, 45(5):1024-1032.

Sangwan, P., Kaushik, K. and Garg, V.K. 2008. 'Vermiconversion of industrial sludge for recycling the nutrients', Bioresour. Technol., 99: 8699-8704.

Sankat Clement K., Castaigne Francois and Maharaj Rohanie. 1996. The air drying behaviour of fresh and osmotically dehydrated banana slices. International Journal of Food Science and Technology, 31(2):123-135.

Sankat, K. Clement. And Castaigne, Francois. 2004. Foaming and drying behaviour of ripe bananas. LWT-Food Science and Technology, 37(5): 517-525.

Sano,H., S. Seo, E. Orudgev, S. Youssefian, K. Ishizuka, Y. Ohashi. 1994. Expression of the gene for a small GTP binding protein in transgenic tobacco elevates endogenous cytokinin levels, abnormally induces salicylic acidacid in response to wounding and increases resistance to tobacco mosaic virus infection, Proc. Natl. Acad. Sci. USA91:10556–10560.

Sanwal, G.G. and Payasi, Anurag. 2007. Garlic extract plus sodium metabisulphite enhances shelf life of ripe banana fruit. International Journal of Food Science and Technology, 42:303-311.

Saravana Kumar, R. and Manimegalai, G. 200I. Storage stability of mixed fruit juice RTS beverages in different storage conditions. Beverage and Food World, 28(2): 28-29.

Saravanan K, Aradhya SM. 2011. Polyphenols of Pseudostem of Different Banana Cultivars and Their Antioxidant Activities. J Agric Food Chem. 2011 Mar 15. [Epub ahead of print]. PMID: 21405133.

Sarkar Supriya, Girisham, S and Reddy, S.M. 2013. Some New Post-Harvest Fungal Diseases of Banana. Annals of Plant Sciences, 02 (05):149-152.

Sarkar, H.N., Hasan, M.A. and Chattopadhyay, P.K. 1995. Studies on shelf life of banana as influenced by chemicals. Journal of Tropical Agriculture. 33(1):97-100.

Sarkar, H.N., Hasan, M.A. and Chattopadhyay, P.K. 1997. influence of polyethylene packing on postharvest storage behaviour of banana fruit. Horticultural Journal. 10(1): 31-39.

Sarrwy, S.M.A., Mostafa, E.A.M. and Hassan, H.S.A. 2012. Growth, Yield and Fruit Quality of Williams Banana as Affected by Different Planting Distances. International Journal of Agricultural Research 7(5):266-275.

Sastry Jayanty, Song, J., Rubistein, Nicole M., Chong Andres., Beaudry, R.M. 2002.Temporal Relationship between Ester Biosynthesis and Ripening Events in Bananas. HortScience., 127:892-1024.

Satyan, S., Scott, K.Y. and Graham, D. 1992. Storage of banana bunches in sealed polyethylene tubes. J. Hort. Sci., 67(2):283-287.

Saucedo-Pompa, S., Jasso-Cantu, D., Ventura-Sobrevilla, J., Saenz-Galindo, A., Rodriguez-Herrera, R. and Aguilar, C.N. 2007. Effect of candelillawax with natural antioxidants on the shelf life quality of freshcut fruits. Journal of Food Quality 30(5):823-836.

Savastano, Jr, H., Warden, P. G. and Coutts, R.S.P. 2003. Potential of Alternative Fiber Cement as Building Material for Developing Area, Journal of Cement and Concrete Composites, 25:586-592.

Savastano, Jr, H., Warden, P.G. and Coutts, R.S.P. 2005. Microstructure and Mechanical Properties of Waste Fiber Cement-Composites, Journal of Cement and Concrete Composites, 27:583-592.

Scalbert A, Manach C, Morand C. 2005. Dietary polyphenols and the prevention of diseases. Crit Rev Food Sci Nutr 45:287–306.

Scott, K.J. and Soertini Gandanegara.1974. Effect of temperature on the storage life of bananas held in polyethylene bags with ethylene absorbent. Trop. Agric., 51(1):23-26.

Seshadri, K., Usman, K. M., Parambaramani, C. 1980. Control of post harvest decay of Robusta banana with an emulsified mineral oil. National Seminar on Banana Production Technology, pp 170-172.

Sevillano, L., Sanchez-Ballesta M.T., Romojaro, F. and Flores, F.B. 2009. Physiological, hormonal and molecular mechanisms regulating chilling injury in horticultural species. Postharvest technologies applied to reduce its impact. J. Sci. Food Agric., 89:555–73.

Seymour, G.B. 1993. Banana. pp 83-106. *In:* Seymour, G.B., J.E. Taylor and G.A. Tucker (eds). Biochemistry of Fruit Ripening. Chapman and Hall, London.

Seymour, G.B., Thompson, A.K., John, P. 1987. Inhibition of degreening in the peel of banana ripened at tropical temperatures. Annu. Appl. Biol. 110:145-151.

Seymour, G.B. 1993. Banana. *In:* Biochemistry of fruit ripening. Seymour, G.B., Taylor, J.E., and Tucker, G.A. (Eds.), Chapman and Hall, 2-6 Boundary Row, London, pp.83-101.

Seymour, G.B., Colquhoun, I.J., DuPont, M.S., Parsley, K.R. and Selvendran, R.R. 1990 Composition and structural features of cell wall polysaccharides from tomato fruits. Phytochem., 29:725–731.

Shaaban, E.A. 1988. Effect of topical application of ethephon on the ripening of banana fruits. Assiut- Journal of Agricultural Sciences. 19(2): 235-245.

Shahidi, F., Naczk, M. 2004. Phenolics in food and nutraceuticals. Boca Raton, FL: CRC Press.

Shantha Krishnamurthy and Krishnamurthy, S. 1989. Storage life and quality of Robusta banana in relation to their stage of maturity and storage temperature. Journal of Food Science and Technology, 26(2): 87-89.

Sharon, N., Lis, H. 1989. Lectins as cell recognition molecules. Science. 246:277–234.

Shaun, R. and Ferris, B. 1997. Improving storage life of plantain banana. IITA research guide No. 62.

Shere, D.M., Kotapalle, S.L. and Kulkarni, D.N. 1993. Effect of maturity stage on qualities of fried banana chips. Journal of Maharashtra Agricultural Universities, 18(2): 338.

Sheshadri, K., Usman, K.M. and Parambaramani, C. 1980. Control of postharvest decay of Robusta banana with an emulsified mineral oil. Proceedings of National Seminar on banana production technology, 170-172.

Shewfelt, R.L. and Prussia, S.E.1993. Postharvest handling, a systems approach. Academic press Limited, USA. pp. 358.

Shewfelt, R.L. 1987. Quality of minimally processed fruits and vegetables. J. Food Quality, 10:143-156.

Sim, C.A. and Bates, R.P. 1994. Challenges to processing tropical fruit juices: Banana as an example. Proceedings of Florida State Horticultural Society, 107:315-319.

Simmonds, N.W. 1959. Bananas. Webster printing service Ltd Bristol. Imperial College of Tropical Agriculture. Trop. Agric. 453-461.

Simmonds, N.W. 1962. The evolution of bananas. Longman, Green and Co.Ltd, London, UK., pp 170.

Simmonds, N.W. 1966. Bananas, 2nd Edn., Longmans Group Ltd. London.

Singh S., Jain S., Singh S.P. and Singh D. 2009. Jams From Combinations Of Different Fruit Pulps. Journal of Food Processing and Preservation (Supplementary), 33(S1): 41-57.

Singh S., Ram H.B. and Tripathi U.K. 1980. Biochemical studies on the developing and ripening of banana, Prog. Hort. 12: 51–57.

Singh Sunita., Trupti Dekhne., Kiran Singh., Kulkarni S.D. 2009. Alcoholic banana beverage aspects in fermentative production. Journal of Food Processing and Preservation, 33(3): 312-329.

Singh, H.P. and Uma, S. 2000. Genetic diversity of Banana in India. Banana Improvement, Production and Utilization (Eds. Singh, H.P and Chadha, K.L.), pp.503.

Singh, R.K. 1976. Time of shooting (flowering) and related yield of banana var. Alpan and Malbhog. Proc. Bihar Acad.Agric.Sci., 24(2):39-44.

Singh, R.K. and Bhattacharya, R.K.1991. Comparative studies on the harvest indices of the leading banana cultivars of North East., Banana Newsletter, 14: 10-11.

Sisler, E.C., Serek, M., Dupille, E. and Goren, R. 1999. Inhibition of ethylene responses by 1-methylcyclopropene and 3-methylcyclopropene. Plant Grow Regul., 27: 105-111.

Smith, C.J.S., Watson, C.F., Bird, C.R., Ray, J., Schuch, W. and Grierson, D. 1990. Expression of a truncated tomato polygalacturonase gene inhibits expression of the endogenous gene in transgenic plants. MOI. Gen. Genet. 224:447-481.

Sole, P. 1996. Banana (Processed). *In:* Processing fruits – Sience and Technology Vol.2, (Es. Somogyi, L.P, BarrettD.M. and Hui, Y.H.), Lancaster, PA Technomic Pub Co.Inc., p.367.

Soltani, M., Alimardani, R. and Omid, M. 2010. Prediction of banana quality during ripening stage using capacitance sensing system. AJCS 4(6):443-447.

Song, J., Rubistein, N. and Beaudry, R.M. 1997. The Temporal Relationship between Volatile Biosynthesis and Other Ripening Parameters in Banana Fruit. HortScience, 32:427-558.

Sreekantaiah, K.R., Jaleel, S.A. and Ramachandra Rao, T.N. 1971. Development studies on enzymatic processing ofbanana J. Fd. Sci. Technol., 8(4):201.

Srivastava, M.K. and Dwivedi, U.N. 2000. Delayed ripening of banana fruit by salicylic acid. Plant Science, 158:87–96.

Srivastava, R.P. 1964. Effect of micro-elements Cu, Zn, Mo, B and Mn on the growth characteristics of banana. Sci. Culture., 30:352-355.

Stevenson, D. 1976. What colour to select for banana bunch covers. Banana Bulletin, 40(4):2.

Steward, F.C. 1960. Physiological investigations on the banana plant: With seven figures in the text. Ann. Bot. 24:117-146.

Stover, R.H. 1979. Pseudostem growth, leaf production and flower initiation in the Grand Naine banana. SIATA Bulletin No.8. La Lima, Honduras, pp.3-37.

Stover, R.H. and Simmonds, N.W. 1987. Banana. Third edition. John Wiley and Sons, New York.

Sulali Anthony, Krishanthi Abeywickrama, Ranjith Dayananada, Shant Wijeratnam, Luxshmi Arambewela. 2004. Fungal pathogens associated with banana, Mycopathologia, 157: 91-97.

Suntharalingam, S. and Ravindran, G.1993. Physical and biochemical properties of green banana flour. Plant Foods for Human Nutrition, 43:19-27.

Suresh, E.R. and Ethiraj, S. 1991. Utilization of overripe bananas for vinegar production. Tropical Science. 31(3): 317-320.

Suvittawat, K and Babpraserth, C. 1996. Banana cultivars for banana chip production. Abstract of International Symposium on Technological Advancements in banana/plantain production and processing. India. pp 47.

Suzanne Sharrock. 1996. Uses of Musa. INIBAP Annual Report. 42-44.

Tapre A.R. and Jain R.K. 2012. Study of advanced maturity stages of banana. IJAERS, 1(3): 272-274.

Tarnowski, Tara L., Jose M. Perez-martinez and Randy C. Ploetz. 2010. Fuzzy Pedicel: A new postharvest disease of banana. Plant Disease, 94(5): 621-627.

Terra, N.N., Garcia, E. and Lajolo, F.M.1983. Starch-Sugar Transformation During Banana Ripening: The behavior of UDP Glucose pyrophosphorylase, Sucrose synthetase and Invertase, J Food Sci, 48:1097-1100.

Tewari, H.K; Marwaha, S.S. and Rupal, K.1986. Ethanol from banana peels. Agricultural Wastes. 16(2):135-146.

Thangaselvabai, T., Suresh, S., Joshua, J. Prem and Sudha, K.R. 2009. Banana Nutrition – A Review. Agric. Rev., 30(1):24-31.

Thangavelu, R., Sundararaju, P. and Sathiamoorthy, S. 2004. Management of anthracnose disease of banana caused by *Colletotrichum musae* using plant extracts. Journal of Horticultural Science and Biotechnology, 79(4):664-668.

Thomas, A.S. 1940. Bananas. III Uganda banana varieties and their uses. *In:* Agriculture in Uganda-Tothik, J.D. Uganda Department of Agriculture. 116-120.

Thomas, P. and Janave, M.T. 1992. Effect of temperature on chlorophyllase activity, chlorophyll degradation and carotenoids of Cavendish bananas during ripening. Int. J. Food Sci. and Technol., 27: 57-63.

Thompson, A.K. and Burden, O.J. 1995. Harvesting and fruit care. *In:* Bananas and Plantains. S.Gowen (Ed.), Chapman and Hall, London. pp.403-432.

Thuwapanichayanan Ratiya; Prachayawarakorn Somkiat; Soponronnarit Somchart. 2012. Effects of Foaming Agents and Foam Density on Drying Characteristics and Textural Property of Banana Foams. LWT Food Science and Technology, 47(2): 348-357.

Toivonen, P.M.A. and C. Lu. 2006. Compositions and methods to improve the storage quality of packaged plants. U.S. Patent Application 20060154822, 13 July 2006.

Toraskar, M.V. and Modi, V.V. 1984. Peroxidase and chilling injury in banana fruit. Journal of Agriculture and Food Chemistry. 32(6): 1352-1354.

Torre-Gutie´rrez La´zaro de la, Chel-Guerrero Luis A., Betancur-Ancona David . 2008. Functional properties of square banana (Musa balbisiana) starch, Food Chemistry, 106:1138–1144.

Totte A, Diaz A, Marouzec, Raoult-Wack AL. 1996. Deep-fat frying of plantain (Musa paradisiaca L.). II. Experimental study of solid/liquid phase contacting systems. Lebensm-Wiss u-Technol 29:599-605.

Toyokuni, S.1999. Reactive oxygen species–induced molecular damage and its application in pathology. Pathol Int, 49:91-102.

Trainotti, Livio, Tadiello Alice and Casadoro Giorgio. 2007. The involvement of auxin in the ripening of climacteric fruits comes of age: the hormone plays a role of its own and has an intense interplay with ethylene in ripening peaches. Journal of Experimental Botany, 58(2): 3299-3308.

Truswell, A.S. 1992. Glycaemic index of foods. European Journal of Clinical Nutrition 46(S1): S91–S101.

Tryford, I.T. 1967. Nutrition, a review of principles and practices. J.Sci Food Agric., 18: 177-183.

Tsen, Jen-Homg., Lin, Yeu-Pyng and King, V. An-Erl. 2003. Banana puree fermentation by Lactobacillus acidophilus immobilization in Ca-alginate. J. Gen. Appl. Microbiol, 49:357-361.

Tucker, G.A. 1993. Introduction. IN: Seymour GB, Taylor JE, Tucker GA (eds). Biochemistry of fruit ripening. Chapman and Hall, London. pp 1-52.

Turner, D.W. 2001. Bananas and Plantains. *In:* Mitra, S.K. (ed.). Postharvest physiology and storage of tropical and subtropical fruit. CAB International. London, UK. pp 47-84.

Turner, D.W. 1995. The response of the plant to the environment. In: Bananas and Plantains. (Ed. S.Gowen). Published in 1995by Chapman & Hall, 2-6 Boundary Row, London SE1 8HN, pp-207-229.

Turner, D.W., Lahav, E. 1983. The growth of banana plants in relation to temperature. Australian Journal of Plant Physiology, 10:43-54.

Tushemereirwe, W.K., Kangire, A., Smith, J., Ssekiwoko, F., Nakyanzi, M., Kataama, D., Musiitwa, C., and Karyaija, R. 2003. An outbreak of bacterial wilt on banana in Uganda. Infomusa, 12(2):6-8.

Twyford, I.T. 1967. Nutrition- A review of principles and practices. J. Sci. Fd. Agric., 18:177-183.

Twyford, I.T. and Walmsley, D. 1968. The status of some micronutrients in healthy Robusta banana plants. Trop. Agric. 45:307-315.

Twyford, I.T. and Walmsley, D. 1974. Uptake and distribution of mineral constituents. Pl. Soil. 41:459-470.

Ukkuru, P.M. and George, D. 1996. Organoleptic and qualitative evaluation of osmotically dried plantain products (Musa, AAB group, Palayankodan). Abstracts of International Symposium on Technological Advancement in banana/plantain Production and Processing. 1996. pp45.

Uma, S., Kalpana, S. and Sathiamoorthy, S. 2003. Banana Fibre. National Research Centre for Banan, Tiruchirapalli, Tamil Nadu, India.

Uma, S., Sathiamoorthy, S and Durai, P. 2005. Banana- Indian Genetic Resources and Catalogue, National Research Centre for Banana (ICAR), Tiruchirapalli – 620 012. India.

Uma, S., Singh, H.P., Dayarani, M. and Shyam, B. 1999. Varietal evaluation for response to edaphic factors on yield and physico-chemcial characteristics of commercial banana cultivars.

Umana-Rojas, G., Garcia, J and Arauz, L.F. 2011. Influence of production system and weather variables in the incidence and severity of crown rot of bananas. Acta Horticulturae, 906.Pp.197-204.

Usherwood, N.R. 1985. In: Potassium in Agriculture. (Munson, R.S. ed.). ASA-CSSA-SSSA, Madison, WI, pp. 489-513.

Vadivel, E. and Shanmugavelu, K.G. 1978. Effect of increasing rates of potash on the quality of banana cv.'Robusta', Potash Rev., 24:1-4.

Van Asten P.J.A., Florent D., Apio M.S. 2010. Opportunities and Constraints for Dried Dessert banana (Musa Spp.) Export in Uganda. Acta Horticulturae No.: 879.

Van der ploeg, A and Heuvelink, E. 2005. Influence of sub-optimal temperature on tomato growth and yield: a Review, The Journal of Horticultural Science & Biotechnology, Vol.80(6): 652-659.

van Iersel, M. F. M., van Dieren, B., Rombouts, F. M. and Abee, T. 1999. Enzyme and Microbial Technology, 24:407–411.

Vatanasuchart Nednapis, Niyomwit Boonma and Wongkrajang Karuna.2012. Resistant starch content, in vitro starch digestibility and physico-chemical properties of flour and starch from Thai bananas, Maejo Int. J. Sci. Technol. 6(02):259-271.

Venkata Subbaiah, K., Jagadeesh, S.L., Sathyanarayana Reddy and Kanamadi, V.C. 2013. Effect of varieties and pre-treatments on physico-chemical and sensory quality of banana papads. Karnataka J. Agric. Sci., 26(1):115-118.

Venkateshwaran, N. and Elayaperumal, A. 2010. Banana fiber reinforced polymer composite- A Review. Journal of Reinforced Plastics and Composites, 29(15): 2387-2396.

Verheul, M.J. 2012. Effects of plant density, leaf removal and light intensity on tomato quality and yield. Acta Horticulturae No.956:365-372.

Vilas-Boas, E.V.de B. and A.A. Kader. 2001. Effect of 1-MCP on fresh-cut fruits. Perishables Handling Quarterly, 108:25.

Vilas-Boas, E.V.de B. and A.A. Kader. 2006. Effect of atmospheric modification, 1-MCP and chemicals on the quality of fresh-cut banana. Postharvest Biol. Technol. 39:155–162.

Villa-Caballero, L., Nava A, Frati, A. and Ponce H. 1999. El estres oxidativo. Es necesario medirlo en el paciente diabetico? Gac Med Mex 136:249–255.

Viquez, F., Lastreto, C. And Cooke, R.D. 1981. A study of the production of clari. ed banana juice using pectinolytic enzymes, Journal of Food Technology, 16:115-125.

Viviani, L. and Leal, P. M. 2007. Revista Brasileira de Fruticultura, 29(3):465-470.

Vojdani, F. and Torres, A. 1990. Potassium sorbate permeability of methylcellulose and hydroxypropyl methylcellulose coatings: Effect of fatty acids. Journal of Food Science, 55:841–846.

Von Uexkll, H.R. 1985. *In:* Potassium in Agriculture. (Munson, R.S. ed.). ASA-CSSA-SSSA, Madison. WI, pp. 929-954.

Vorm, P.D.J. van der and Diest, A.van. 1982. Redistribution of nutritive elements in a 'Gros Michel' banana plant. Neth. J. Agric. Scie., 30:286-296.

Wade, N. L., Satyan, S. H. and Kavanagh, E.E. 1993. Increase in lowmolecular size uronic-acid in ripening banana fruit. Journal of the Science of Food and Agriculture, 63:257–259.

Wakabayashi, K., 2000. Minireview: changes in cell wall polysaccharides during fruit ripening. J. Plant Res. 113:231-237.

Wallner, S.J., Bloom, H.L. Characteristics of tomato cell wall degradation in vitro: implications for the study of fruit-softening enzymes. Plant Physiol. 1977 Aug;60(2):207–210.

Walmsley, D. and Twyford, I.T. 1975. Mineral composition of banana plant.5. Sulphur, iron, manganese, boron, zinc, copper, sodium and aluminum. Pl. Soil., 45: 595-611.

Wang Juan, Li Zhi Yuan, Chen Ren Ren, Bao Yong Jin adn Yang Ming Gong. 2007. Comparison of volatiles of banana powder dehydrated by vacuum belt drying, freeze-drying and air-drying. Food Chemistry 104:1516–1521.

Wang, Y., Lu, W.J., Jiang, Y.M., Luo, Y.B., Jiang, W.B. and Joyce, D. 2006. Expression of ethylene-related expansin genes in cool-stored ripening banana fruit. Plant Sci., 170:962–967.

Waskar, D.P. and S.K. Roy 1993. Effect of Zero Energy cool chamber on storage of banana. Maharashtra J. Hart. 7:37-45.

Watada, A.E., Abe, K. and Yamauchi, N.1990. Physiological activities of partially processed fruits and vegetables. Food Technol. 44(5):116-122.

Watada, A.E., Abe, K. and Yamuch, N. 1986. Physiological activity of partially processed fruits and vegetables. Food Technology. 40:82-85.

Watada, A.E.; Ko, N.P. and Minott, D.A. 1996. Factors affecting quality of fresh-cut horticultural products. Postharvest Biol. Technol. 9:115-125.

Waterhouse, D.F. and Norris, K.R. 1987. Biological Control Pacific Prospects. Melbourne (Inkata Press): VIII + 454 S.; geb. 130. Austr. Doll. . Dtsch. Entomol. Z., 37: 360.

Weber, N.D., Andersen, D.O., North, J.A., Murray, B.K., Lawson, L.D. and Hughes, B.G. 1992 *In vitro* virucidal effects of *Allium sativum* (garlic) extractand compounds. Planta Medica. 58:417-417.

Weerasinghe, P. and Premalal, N.H.R. 2002. Influence of potassium fertilization of growth and yield of Embul Banana (Musa spp. AAB Group) grown in Rhodudalfs under irrigated conditions. Annal. Sri Lanka Dep. Agric., 4:109-117.

Whitney SEC, Gidley MJ and McQueen-Mason SJ 2000. Probing expansin action using cellulose/hemicellulose composites. The Plant J. 22: 327-334.

Willaert, R., and Nedovic, V.A. 2006. Journal of Chemical Technology and Biotechnology (Oxford, Oxfordshire), 81:1353–1367.

Wills, R., McGlasson, B., Graham, D. and Joyce, D., 1998. Postharvest: an Introduction to the Physiology and Handling of Fruit, Vegetables and Ornamentals. University of New South Wales Press Ltd., Sydney, Australia.

Wilson Wijerathnam, R.S., Jayatilake, S., Hewage, S.K., Perera, L.R., Paranerupasingham, S. and Peiris, C.N. 1994. *In:* Postharvest Handling of Tropical Fruits (Eds): Champ, B.R; Highley, E. and Johnson, G.I. ACIAR, Canberra (Australia).

Wong, D.W.S., Tillin, S.J., Hudson, J.S., and Pavlath, A.E. 1994. Gas exchange in cut apples with bilayer coatings. Journal of Agricultural Food Chemistry, 42:2278–2285.

Yadav, B.Ritika., Yadav, S.Baljeet and Kalia Navneet, 2010. Development and storage studies on whey-based banana herbal (*Mentha arvensis*) beverage. American Journal of Food Technology, 5(2): 121-129.

Yadav, I.S. *et al.*, 1988. South Indian Hort., 36:167-169.

Yadlod, S.S. and Kadam, B.A. 2008. Effect of Plant Growth Regulators And Micronutrients On Growth, Yield, And Storage Life of Banana (*Musa* Spp) Cv. Ardhapuri. Agric. Sci. Digest, 28(4):304-306.

Yadlod, S.S. and Kadam, B.A. 2008. Effect of Plant Growth Regulators and Micronutrients on growth, yield, and storage life of banana (*Musa* Spp) Cv. Ardhapuri, Agric. Sci. Digest, 28(4):304-306.

Yang, C., Z. Nong, J. Lu, L. Lu, J. Xu, Y. Han, Y. Li and S. Fujita, 2004. Banana polyphenol oxidase: Occurrence and change of polyphenol oxidase activity in some banana cultivars during fruit development. Food Sci. Technol. Res., 10:75-78.

Yoshioka, H., Ueda, Y. and Chachin, K. 1980b. Effect of high temperature (40°C) on changes in acid-phosphatase activity during ripening in banana. Journal of Japanese Society of Food Science and Technology. 27(10):511-516.

Yoshioka, H., Ueda, Y. and Iwata, T. 1982. Development of isoamyl acetate biosynthetic pathway in banana fruits during ripening and suppression of its development at high temperature. Journal of Japanese Society of Food Science and Technology. 29(6):333-339.

Yoshioka, H., Ueda, Y. and Chachin, K. 1980a. Inhibition of banana fruit ripening and decrease of protein synthesis at high temperature (40°C) Physiological studies of fruit ripening in relation to heat injury. Part: III. Journal of Japanese Society of Food Science and Technology. 27(12):610-615.

Zawistowshy, J, Biliaderis, C.G. and N. A. M. Eskin. 1991. Polyphenol Oxidase. *In:* D. S. Robinson and N. A. M. Eskin, Eds., Oxidative Enzymes in Foods, Elsevier, London, pp. 217-273.

Zewter, A., Woldetsadik, K. and Workneh, T.S. 2012. Effect of 1-methylcyclopropene, potassium permanganate and packaging on quality of banana. African Journal of Agricultural Research, 7(16):2425-2437.

Zhang, P., Whistler, R.L., BeMiller, J.N. and Hamaker, B.R. 2005. Carbohydrate Polymers, 59:443–458. 19.

Zocchi, G. and Mignani, I. 1995. Calcium physiology and metabolism in fruit trees. Acta Hort. 383: 15-23.

Zeitfracht Medien GmbH
Ferdinand-Jühlke-Straße 7
99095 Erfurt, Deutschland
produktsicherheit@kolibri360.de